Communications and Control Engineering

Springer
London
Berlin
Heidelberg
New York
Barcelona
Hong Kong
Milan
Paris
Singapore
Tokyo

Published titles include:

Nonlinear Control Systems (3rd edition)
Alberto Isidori

Theory of Robot Control
C. Canudas de Wit, B. Siciliano and G. Bastin (Eds)

Fundamental Limitations in Filtering and Control
María M. Seron, Julio Braslavsky and Graham C. Goodwin

Constructive Nonlinear Control
R. Sepulchre, M. Janković and P.V. Kokotović

A Theory of Learning and Generalization
M. Vidyasagar

Adaptive Control
I.D. Landau, R. Lozano and M.M'Saad

Stabilization of Nonlinear Uncertain Systems
Miroslav Krstić and Hua Deng

Passivity-based Control of Euler-Lagrange Systems
Romeo Ortega, Antonio Loría, Per Johan Nicklasson and Hebertt Sira-Ramírez

Stability and Stabilization of Infinite Dimensional Systems with Applications
Zheng-Hua Luo, Bao-Zhu Guo and Omer Morgul

Nonsmooth Mechanics (2nd edition)
Bernard Brogliato

Nonlinear Control Systems II
Alberto Isidori

L_2-Gain and Passivity Techniques in Nonlinear Control
Arjan van der Schaft

Control of Linear Systems with Regulation and Input Constraints
Ali Saberi, Anton A. Stoorvogel and Peddapullaiah Sannuti

Robust and $H\infty$ Control
Ben M. Chen

Computer Controlled Systems
Efim N. Rosenwasser and Bernhard P. Lampe

Dissipative Systems Analysis and Control
Rogelio Lozano, Bernard Brogliato, Olav Egeland and Bernhard Maschke

Control of Complex and Uncertain Systems
Stanislav V. Emelyanov and Sergey K. Korovin

Goro Obinata and Brian D.O. Anderson

Model Reduction for Control System Design

With 39 Figures

Springer

Goro Obinata, PhD
Dept. Mechanical Engineering, Faculty of Engineering and Resource Science, Akita University, 1-1 Tegatagakuen University, Akita City, 010-8502 Japan

Brian D.O. Anderson, PhD
Research School of Information Sciences and Engineering, Australian National University, Canberra ACT 0200, Australia

Series Editors
E.D. Sontag • M. Thoma

ISSN 0178-5354

ISBN 1-85233-371-5 Springer-Verlag London Berlin Heidelberg

British Library Cataloguing in Publication Data
A catalog record for this book is available from the British Library

Library of Congress Cataloging-in-Publication Data
Obinata, Goro, 1949-
Model reduction for control systems design / Goro Obinata and Brian D.O. Anderson.
p. cm. -- (Communications and control engineering, ISSN 0178-5354)
ISBN 1-85233-371-5 (alk. paper)
1. Automatic control. 2. Control theory. I. Anderson, Brian D.O. II. Title. III. Series.
TJ213 O15 2000
629.8--dc21 00-061254

Typesetting: Camera ready by authors
Printed and bound by Athenæum Press Ltd., Gateshead, Tyne & Wear
69/3830-543210 Printed on acid-free paper SPIN 10775966

This book first appeared in Japanese in December 1999 with the title Seigyo system sekkei controller no teijigenka, as Volume 21 in the series System Control and Information Library. The English edition appears with the permission of the Japanese publisher Asakura Shoten.

Preface

The aim of this book is to provide the theoretical foundations of model and controller reduction for students and industrial research and development engineers, especially for those concerned with the application of advanced control theory to actual systems. Readers will obtain a good perspective of recent developments on these problems, and also a grounding in design techniques for practical applications.

This book has its roots in work going back several decades. Modern methods for model reduction perhaps started with a 1966 paper of E. J. Davison. The key feature of the technique advanced in the paper was to retain the dominant modes of the original system and to discard the nondominant modes. This idea is almost the same as a method long used by control engineers, termed dominant pole approximation; however, it was the modern formulation of Davison's paper which stimulated many researchers to move into the field, resulting in an explosion in the development of techniques. In the early work, methods for approximating typical responses of high order were sought. Many papers were published on topics such as minimization of L_2 impulse response error norm, continued fraction approximation, and so on. Then in 1981, B. C. Moore, seeking to introduce the principal component method of statistical analysis into linear dynamical systems, advanced the concept of balanced realizations. This was a major step forward, allowing as it did the introduction of the so-called balanced truncation method for reducing a stable linear system with an associated *a priori* error bound. Some of these reduction schemes are reviewed in Chapter 1 of this book.

Simultaneously with the work on model reduction, work on linear-quadratic-gaussian (LQG) design methods had been progressing and then, beginning in the 1980s, came the developments in H_∞ control design, μ-synthesis and linear matrix inequalities (LMIs). Especially with the latter, it seemed that nearly all linear controller design problems could now be solved. One problem however remained difficult: design of low order controllers for high order plants. The advanced controller design methods tend to supply controllers with order comparable to the plant order, and therefore controllers often of high order. Intuitive understanding of how the controller is functioning and actual implementation in a reliable manner are major tasks with a high order controller, and so a demand arose for techniques of controller order reduction. This cannot be straightforwardly delivered by LMI methods.

The available methods of model reduction did not prove of themselves an effective technique for controller reduction either, since they provide no straightforward way for handling the problem of retaining closed-loop stability and performance when the reduced order controller replaces the full order controller. A moment's reflection shows that if these problems are to be properly addressed, the plant itself has to somehow be brought into the reduction task.

Perhaps the first time this occurred was in a key development of the balanced truncation idea, due to D. F. Enns in 1984 who showed how to capture the plant into the controller reduction problem. His method relied on introducing frequency weighting into the balanced truncation procedure. Controller reduction is the major theme of Chapter 3, when Enns' scheme and other schemes are explored. Another major advance in the model reduction area had come with a famous paper of K. Glover which also appeared in 1984. This paper showed how with state-variable type calculations, a series of approximation problems involving optimal approximation with an unusual norm, the Hankel norm, could be exactly solved. The solutions were also relevant to approximation with more conventional error measures than the Hankel norm. In due course, it was seen how frequency weighting could be introduced into this problem also, so that it could be used as a basis for controller reduction.

The work of Enns and Glover with or without frequency weighting was principally concerned with minimizing a so-called additive error, measured for scalar transfer functions by the difference between two corresponding points on their Nyquist diagrams. Another common form of error between transfer functions can be obtained using the distance between corresponding points on Bode diagrams of amplitude and phase. Focusing on this type of error gives rise to multiplicative or relative error approximation problems, and analogs of balanced truncation approximation and optimal Hankel norm approximation became available for this type of error, starting around the mid 1980s and continuing for almost ten years. K. Glover, M. Green and others developed this type of model reduction, which is described in detail in Chapter 2.

As a further major direction of development occurring principally in the 1990s, we cite the application of coprime fractional descriptions, especially in controller reduction. In Chapter 4 of this book, some techniques of this category are explained, preceded by a short introduction to coprime fractional descriptions.

Acknowledgements

The writing of this book was proposed in 1997 by Professor Akira Osumi, of Kyoto Institute of Technology, for one of a Japanese series called the Systems, Control and Information Library, a series sponsored by the Japanese Society for Systems, Control and Information. At that time, Goro Obinata proposed the writing collaboration with Brian Anderson. The writing was carried out in Japanese and English simultaneously, in Akita, Japan and in Canberra, Australia and in two one-week collaborations on the Australian Pacific coast.

Finally, we would like to thank Astushi Nakayama for solving the examples, typing with TEX and making the corrections to the Japanese version through the careful reading of the draft, Yoshikazu Kobayashi for solving the examples, and Kiyoshi Nagasaku for preparing the figures. The work of Jenni Watkins on the English language version is also gratefully acknowledged. James Ashton assisted with Japanese/English electronic interfaces. We would also like to acknowledge Professor Osumi for proposing this book, and the editorial staff of Asakura Shoten for their assistance in publishing the Japanese edition, which appeared in December 1999.

Goro Obinata and Brian D. O. Anderson

Contents

List of Figures

Key Notation

A^T	Transpose of A.
A^*	Complex conjugate transpose of A.
A^+	Pseudo-inverse of A.
$\lambda_i(A)$	Eigenvalue of A.
$\lambda_{\max}(A)$	Eigenvalue of A of maximum modulus.
$\bar{\sigma}(A)$	Maximum singular value of A.
$\sigma_i(G)$	ith Hankel singular value of a stable transfer function matrix $G(s)$.
$G_*(s)$	$G^T(-s)$.
$[G(s)]_+$	Proper part of the strictly stable part of $G(s)$ in an additive decomposition, *i.e.*, the sum of the strictly proper partial fraction summands with poles in the open left half plane plus $G(\infty)$.
$\mathcal{F}_l(\ ,\ )$	Lower Linear Fractional Transformation (explained in text).
RH_∞	Set of stable rational proper transfer function matrices.
RH_∞^-	Set of antistable rational proper transfer function matrices.
$RH_\infty^-(r)$	Set of rational proper transfer function matrices $X(s)$ such that $[X(s)]_+$ has degree at most r.
$S(n_1, n_2)$	Set of rational proper transfer function matrices with n_1 poles in $\mathrm{Re}[s] < 0$ and n_2 poles in $\mathrm{Re}[s] \geq 0$.

Chapter 1

Methods for Model Reduction

1.1 Introduction to Model and Controller Reduction

The aim of this book is to introduce the reader to problems of model and controller reduction, with an emphasis on the latter. The book is restricted to considering linear, time-invariant models and controllers.

Model order reduction and controller order reduction serve different objectives.

Problems of model order reduction can arise for at least two distinct reasons. First, the model in question may be a filter required for signal processing, in which case it is usually stable. One wants to simplify it to allow easier implementation. Second, the model in question may be a model of a plant for which it is intended to perform a controller design (and it may not be stable). For the purposes of controller design, one would like to have as accurate a representation of the plant as possible. Model order reduction, *i.e.*, replacing a high order model of the plant by a low order model, in a sense will only be necessary if the dimensions of an initially given model are so large as to preclude effective calculations, including design of the controller.

On the other hand the retention of a high order model of the plant is one of the reasons for there being a controller reduction problem. While advanced control theory such as H_∞ robust control or μ-synthesis are receiving increasingly wide use in applications (normally with the aid of sophisticated software design tools), the resulting controllers typically have order similar or equal to that of the plant model. Thus a high order plant model can give rise to a high order controller. The practical implementation of such a high order controller is however not necessarily an easy task. There may be pressure to have a lower order for saving hardware resources and avoiding numerical difficulties in the digital processing.

Problems of controller reduction are more difficult than those of model reduction. This is because the controller is part of a closed loop and one wants to reduce so as to cause little change to closed loop behaviour. The difficulties of controller reduction, as opposed to simple model reduction, have been a significant target for researchers of control theory during the last twenty years or so.

In this chapter, various methods for model reduction are introduced. While not necessarily suitable in themselves for controller reduction, many provide a basis for controller reduction methods to be presented subsequently. We defer to the next chapter model reduction methods based on multiplicative or relative error measures.

The methods are largely restricted to transfer function matrices which are rational and stable, *i.e.*, the only singularities are poles, and they lie in $\mathrm{Re}[s] < 0$. The class of such transfer function matrices is termed RH^∞.

Main points of the section

1. Model and controller reduction will often have different purposes. A key purpose of this book is to explain controller reduction.

1.2 Model Reduction by Truncation

Consider the linear system given by

$$\dot{x} = Ax + Bu, \tag{1.2.1}$$

$$y = Cx + Du, \tag{1.2.2}$$

where the system matrices are decomposed as:

$$A = \begin{bmatrix} A_{11} & A_{12} \\ A_{21} & A_{22} \end{bmatrix}, \qquad B = \begin{bmatrix} B_1 \\ B_2 \end{bmatrix}, \qquad C = \begin{bmatrix} C_1 & C_2 \end{bmatrix}. \tag{1.2.3}$$

In this chapter, the initially given system, which is to be reduced, is assumed to be asymptotically stable, controllable, and observable. Using these partitioned matrices, we define the truncated reduced order model as follows:

$$\dot{x}_r = A_{11}x_r + B_1 u, \tag{1.2.4}$$

$$y_r = C_1 x_r + Du. \tag{1.2.5}$$

The order of this model is r with A_{11} an $r \times r$ matrix. The additive error between the rth order model and the original system $G(s) = D + C(sI - A)^{-1}B$ can be given in terms of the difference of their transfer functions:

$$G(s) - G_r(s) = \tilde{C}(s)\Delta^{-1}(s)\tilde{B}(s). \tag{1.2.6}$$

Here,

$$\begin{aligned} G_r(s) &= D + C_1 (sI - A_{11})^{-1} B_1, \\ \Delta(s) &= sI - A_{22} - A_{21}\phi(s)A_{12}, \\ \tilde{B}(s) &= A_{21}\phi(s)B_1 + B_2, \\ \tilde{C}(s) &= C_1\phi(s)A_{12} + C_2, \\ \phi(s) &= (sI - A_{11})^{-1}. \end{aligned} \tag{1.2.7}$$

The calculation is not difficult, but the reader is recommended to try it once. The simplest way is to use the fact that

$$\begin{aligned} sI - A = &\begin{bmatrix} I & 0 \\ -A_{21}\phi(s) & I \end{bmatrix} \\ &\times \begin{bmatrix} sI - A_{11} & 0 \\ 0 & sI - A_{22} - A_{21}\phi(s)A_{12} \end{bmatrix} \begin{bmatrix} I & -\phi(s)A_{12} \\ 0 & I \end{bmatrix}. \end{aligned} \tag{1.2.8}$$

The error is dependent upon the state coordinate basis of the original system. We will mention several methods to select the state coordinate basis (to secure a low error approximation) in later sections. Observe that no matter what the basis, the truncated model has a certain error at $\omega = 0$, in general nonzero, and matches the original system at $\omega = \infty$:

$$G_r(0) \neq G(0), \qquad G_r(\infty) = G(\infty). \tag{1.2.9}$$

The choice of coordinate basis is often recast as a choice of matrices L and R such that

$$A_r = LAR, \qquad B_r = LB, \qquad C_r = CR, \qquad LR = I. \tag{1.2.10}$$

For suppose that T is a nonsingular matrix such that the original A, B, C triple is replaced by $T^{-1}AT, T^{-1}B, CT$ and then truncated. If one identifies L with the first r rows of T^{-1} and R with the first r columns of T, then (1.2.10) results. Conversely, if L and R are known one can easily find augmenting matrices $\bar{L}$ and $\bar{R}$ such that $[L^T \; \bar{L}^T]^T$ and $[R \; \bar{R}]$ are nonsingular and inverses of one another, and then it is evident that $T = [R \; \bar{R}]$.

The truncated model concept will be developed in various ways in later sections.

Mode truncation and aggregation

If the matrix A in (1.2.1) is diagonalized (or put into Jordan canonical form in the case of repeated eigenvalues) by change of coordinate basis, r poles of the truncated model given by (1.2.4) and (1.2.5) are identical to a subset of those of the original system (since $A_{21} = 0$). The reduced order model so obtained has a steady state error to a step input in general, and several techniques have been proposed to modify or to eliminate the steady state error including adjustment of the input and output coupling matrices B_r and C_r, Siret, Michailesco and Bertrand (1977), Bonvin and Mellichamp (1982). In general, one should select the most dominant eigenvalues of A as eigenvalues of A_r for a small reduction error. Of course, the question of how one measures error can be critical. The transfer function $ab(s+a)^{-1}$ has a steady state step response and a maximum imaginary axis gain of b; the L_2 norm is $b(a/2)^{1/2}$; selections of a exist which make either of these measures arbitrarily greater than the other. In fact, selection of modes can be difficult when the fast modes have big gains, or indeed when there are many eigenvalues which are close to each other. Some criteria have been proposed for selecting eigenvalues, *e.g.*, Enright and Kamel (1980).

Aggregation, see *e.g.*, Aoki (1968) is one of the methods which retain certain eigenvalues of the original system. For simplicity, we assume that there are no repeated eigenvalues in A. Let L be a $k \times n$ matrix whose row vectors are A^T-invariant and linearly independent. Then one defines a new state vector z by

$$z(t) = Lx(t). \tag{1.2.11}$$

Because of the A^T invariance property, there exists A_r and one can define B_r such that

$$LA = A_r L, \qquad LB = B_r. \tag{1.2.12}$$

It is easily confirmed that A_r is given explicitly as $A_r = LAL^T(LL^T)^{-1}$. The aggregated model for (1.2.1) is

$$\dot{z} = A_r z + B_r u. \tag{1.2.13}$$

If $x(0) \neq 0$ but $z(0) = Lx(0) = 0$ and $u(t) = 0$ for $t > 0$, then $z(t) = 0$ for all $t > 0$. This means that the pair (A, L) is unobservable. To obtain an output equation for the aggregated model, choose for the output equation $C_r = CL^T(LL^T)^{-1}$. The choices of A_r, B_r and C_r satisfy (1.2.10) when $R = L^T(LL^T)^{-1}$.

Main points of the section

1. A method for reducing the order of models by truncation of state variable realizations is described, and a formula for the additive error associated with a truncated model is given.

2. The transfer functions of the original system and truncated approximation match at $\omega = \infty$.

3. An important special case of truncated models is termed mode truncation, and this also is achieved through state aggregation with approximate definition of the output.

1.3 Singular Perturbation

The singular perturbation technique has a long history in mathematics and indeed in control theory. It is generally thought of as providing a form of approximation based on the rejection of very fast modes from a model. Often, one thinks of the method as being used when the original system is parametrized by a scalar parameter μ (normally assumed to be small), with a description such as

$$\dot{x} = A_{11}x + A_{12}z + B_1u, \tag{1.3.1}$$

$$\mu\dot{z} = A_{21}x + A_{22}z + B_2u, \tag{1.3.2}$$

$$y = C_1x + C_2z + Du. \tag{1.3.3}$$

We can see this kind of system frequently when the overall system contains subsystems of very differing characteristic frequencies. The approximation is obtained by replacing the small parameter μ by zero. Setting $\mu = 0$ in (1.3.2) yields

$$A_{21}x + A_{22}z + B_2u = 0. \tag{1.3.4}$$

Assuming A_{22} is invertible, one can substitute the solution z of (1.3.4) into (1.3.1) and (1.3.3) to obtain

$$\dot{\tilde{x}} = \left(A_{11} - A_{12}A_{22}^{-1}A_{21}\right)\tilde{x} + \left(B_1 - A_{12}A_{22}^{-1}B_2\right)u, \tag{1.3.5}$$

$$y = \left(C_1 - C_2A_{22}^{-1}A_{21}\right)x + \left(D - C_2A_{22}^{-1}B_2\right)u. \tag{1.3.6}$$

Under normal circumstances, the original system is stable for all sufficiently small μ. This ensures that the reduced order system is also stable.

The fact that the reduced order model is obtained from the original model by a limiting process means that many conclusions and calculations (including those reached by for example LQG design methods) can be regarded as limits of the corresponding conclusions or calculations for the original unreduced model. See *e.g.*, Kokotovic, O'Malley and Sannuti (1976). One important property of the reduced order model is that the steady state gain matches that of the original system, which is in contrast to the situation with truncated models, that is,

$$G_r(0) = G(0). \tag{1.3.7}$$

On the other hand, there is normally no match at $\omega = \infty$.

Notice also that if a high order model is presented as in equations (1.3.1) to (1.3.3) but without presence of the parameter μ (*i.e.*, $\mu = 1$), one can still construct the approximation of (1.3.5) and (1.3.6); the approximation makes more sense when a small μ is present, but there is no in-principle objection to making the approximation when $\mu = 1$.

Generalized singular perturbation

We can extend this classical method to a more general one. The generalization gives us in some sense a continuous connection between singular perturbation and truncation.

Using (1.2.4) and (1.2.5), we defined the truncated model and using (1.2.6) and (1.2.7) we defined the error. Now suppose we reverse the roles of reduced order model and error; thus provisionally we regard the reduced order model as $\tilde{C}(s)\Delta^{-1}(s)\tilde{B}(s)$. This is not particularly practical, since $\tilde{B}(s)$ and $\tilde{C}(s)$ are not constant matrices, and Δ^{-1} will not be of the form $(sI - A)^{-1}$ for some low dimension A. These defects can however be remedied. Let ρ be a constant in the closed interval $[0, \infty]$ replacing s in $\phi(s)$, $\tilde{B}(s)$ and $\tilde{C}(s)$. Then we get the generalized perturbation model with adjusted feedthrough term $\tilde{D}(\rho) = C_1(\rho I - A_{11})^{-1}B_1$ as follows:

$$G_r(s) = \tilde{C}(\rho)\tilde{\Delta}^{-1}(s)\tilde{B}(\rho) + \tilde{D}(\rho), \tag{1.3.8}$$

where

$$\begin{aligned}
\tilde{\Delta}(s) &= sI - A_{22} - A_{21}\phi(\rho)A_{12}, \\
\tilde{B}(\rho) &= A_{21}\phi(\rho)B_1 + B_2, \\
\tilde{C}(\rho) &= C_1\phi(\rho)A_{12} + C_2, \\
\phi(\rho) &= (\rho I - A_{11})^{-1}.
\end{aligned} \tag{1.3.9}$$

As a procedure for model approximation, this would seem to be rather speculative. But observe that if we set $\rho = 0$, then (1.3.8) becomes the standard singular perturbation model which approximates the original system in the low frequency region with exact matching at $\omega = 0$. Furthermore, the truncated model $C_2(sI - A_{22})^{-1}B_2$ comes out if we set $\rho = \infty$, and of course the reduced order model approximates the original system well at high frequencies. When ρ is fixed but arbitrary, the transfer function of the reduced order model exactly matches the transfer function of the original system at $s = \rho$. Unfortunately, we cannot choose ρ as an arbitrary imaginary number to match the frequency response to the original system at some point on the imaginary axis, since the resulting reduced order model would fail to be physically realizable through the coefficients in the system matrices becoming complex numbers.

We have said nothing about estimating the errors. It turns out, as explained in the next section, that if we perform singular perturbation (either the normal variety or its generalization) starting with a balanced realization, then we get a pleasing error bound.

Main points of the section

1. In the singular perturbation technique, setting a single parameter multiplying a derivative to zero yields the reduced order model.

2. The frequency domain error between the original system and reduced order model is zero at $\omega = 0$, *i.e.*, the steady-state gain is preserved.

3. Singular perturbation techniques for model reduction can be generalized, and both truncation and singular perturbation can be regarded as particular cases of generalized singular perturbation.

1.4 Reduced Order Models Based on Balanced Realization Truncation

In Moore (1981), the idea of balanced realization of a stable transfer function is derived from the concept of principal component analysis in statistical methods: that is, the controllability grammian and observability grammian are supposed to be decomposed into principal components for evaluating the contributions to the response of each mode. The techniques have found wide acceptance because of their ease of use and the fact that several analytical results and variations of the techniques can be obtained.

The key property of a balanced realization is that the state coordinate basis is selected such that the controllability and observability grammians are both equal to some diagonal matrix, Σ say, normally with the diagonal entries of Σ in descending order. The state space representation is then called a balanced realization.

The magnitudes of the diagonal entries of the grammians reflect the contributions of different entries of the state vector to system responses. More precisely, suppose that the ith diagonal entry of both grammians is σ_i. Consider the task of controlling the system from the origin at time $-\infty$ to the state e_i (unit vector with 1 in the ith position) at time 0. The minimum energy for doing this, as measured by the integral of the control norm squared, is σ_i^{-1}. Thus state vector entries which are little excited by an input are those associated with small σ_i. Consider also the output energy (*i.e.*, integral of the output norm squared, over $[0, \infty]$) resulting from an initial state at $t = 0$ of e_i and zero excitation. This is σ_i. Thus state vector entries which contribute least to the output are associated with small σ_i. All this suggests that the least important entries of the state vector are those associated with the smallest σ_i. Thus, we can use truncation as described in Section 1.2 to eliminate unimportant state variables from the realization. The following two Lyapunov equations give the relation to the system matrices $\bar{A}$, $\bar{B}$, $\bar{C}$ for a balanced realization:

$$\bar{A}\Sigma + \Sigma\bar{A}^T + \bar{B}\bar{B}^T = 0, \tag{1.4.1}$$

$$\Sigma\bar{A} + \bar{A}^T\Sigma + \bar{C}^T\bar{C} = 0. \tag{1.4.2}$$

Hence $\Sigma = \text{diag}[\sigma_i]$ and $\sigma_1 \geq \sigma_2 \geq \cdots \geq \sigma_n$. The σ_i are termed the Hankel singular values (of the original transfer function matrix).

Let us now consider how a balanced realization may be obtained; then we shall address the question of bounding the approximation error resulting from truncation.

Let P and Q be respectively the controllability grammian and observability grammian associated with an arbitrary minimal realization $\{A, B, C\}$ of a stable transfer function, respectively. Since P and Q are symmetric, there exist orthogonal transformations U_c and U_o, such that

$$P = U_c S_c U_c^T, \tag{1.4.3}$$

$$Q = U_0 S_0 U_0^T, \tag{1.4.4}$$

where S_c, S_o are diagonal matrices. The matrix

$$H = S_0^{1/2} U_0^T U_c S_c^{1/2} \tag{1.4.5}$$

is constructed, and a singular value decomposition is obtained from it:

$$H = U_H S_H V_H^T. \tag{1.4.6}$$

Using these matrices, the balancing transformation is given by

$$T = U_0 S_0^{-1/2} U_H S_H^{1/2}. \tag{1.4.7}$$

The balanced realization is $\bar{A} = T^{-1}AT$, $\bar{B} = T^{-1}B$, $\bar{C} = CT$. The following property is easily confirmed through simple manipulations.

$$T^T Q T = T^{-1} P T^{-T} = S_H. \tag{1.4.8}$$

Thus $S_H = \Sigma$.

Unforced state trajectories

In a balanced realization with zero input, it is always true that state trajectories decay in norm, *i.e.*, $x^T(t)x(t)$ is a decaying function. To see this, observe that along unforced trajectories

$$\frac{d}{dt}\left[x^T(t)x(t)\right] = x^T(t)\left[\bar{A}^T + \bar{A}\right]x(t), \tag{1.4.9}$$

while also the two equations satisfied by Σ yield

$$\Sigma(\bar{A}^T + \bar{A}) + (\bar{A}^T + \bar{A})\Sigma = -\bar{B}\bar{B}^T - \bar{C}^T\bar{C}. \tag{1.4.10}$$

From this Lyapunov equation it follows that all eigenvalues of $\bar{A} + \bar{A}^T$ necessarily have nonpositive real parts. Since $\bar{A} + \bar{A}^T$ is symmetric, they either are negative or zero. Since $\bar{A}$ is asymptotically stable, $x^T(t)x(t) \to 0$ as $t \to \infty$ for all $x(0)$. From this it follows easily that $x^T(t)x(t)$ is strictly decreasing for almost all t.

Properties of the truncated system

A truncation of the associated balanced realization amounts to throwing away certain rows and columns of Σ, as well as of $\bar{A}$, $\bar{B}$ and $\bar{C}$. If we write Σ as

$$\Sigma = \begin{bmatrix} \Sigma_1 & 0 \\ 0 & \Sigma_2 \end{bmatrix}, \tag{1.4.11}$$

then after truncation, Σ_1 is left and it is the controllability and observability grammian of the reduced order model—at least if the reduced order model is asymptotically stable. It is normal in truncation to insist that the smallest (last) entry of Σ_1 exceeds the largest (first) entry of Σ_2; more effective truncations in a sense are obtained when the dropped values are much smaller.

Assume that Σ_1 and Σ_2 have no diagonal entries in common. Then the truncated system is provably asymptotically stable, controllable, and observable—the argument is not difficult. [In fact, if the two diagonal matrices Σ_1 and Σ_2 have one or more entries in common, it is not guaranteed that the reduced order system is asymptotically stable, although an adjustment process is available to patch up the problem, Pernebo and Silverman (1982).] Asymptotic stability of the reduced order system is necessary for the interpretation of the truncated Σ_1 matrices to be controllability and observability grammians of the reduced order system.

An error bound for balanced truncation

We turn now to an error bound formula. An infinity norm bound on the absolute error in the frequency domain is provided as follows, see Glover (1984) and Enns (1984):

$$E_\infty = \|G(s) - G_r(s)\|_\infty \leq 2\,\mathrm{tr}\,[\Sigma_2] = 2 \sum_{k=r+1}^{n} \sigma_k. \tag{1.4.12}$$

Actually, in case Σ_2 has repeated entries a tighter bound is available: one computes the right hand side of (1.4.12) only counting a single occurrence of each distinct Hankel singular value. The bound (1.4.12) will be proven by the following two steps. First, we shall give a direct proof of the error bound for the case of $r = n - 1$; then we shall consider the case of general r. In the modification of the bound of (1.4.12) to cope with repeated Hankel singular values, the key is to prove a bound when the last m Hankel singular values are the same and one considers reduction from order n to order $n - m$; we shall omit the details, which are not difficult once the proof for the case of no repeated singular values is understood.

With $r = n - 1$, the quantities $\tilde{B}(s)$ and $\tilde{C}(s)$ appearing in the error formula (1.2.6) are scalar. Accordingly, (1.2.6) on the imaginary axis yields

$$\bar{\sigma}\,[G(j\omega) - G_r(j\omega)] = \lambda_{\max}^{1/2}\big[\Delta^{-1}(j\omega)\tilde{B}(j\omega)\tilde{B}^*(j\omega)(\Delta^{-1}(j\omega))^*\tilde{C}^*(j\omega)\tilde{C}(j\omega)\big]. \tag{1.4.13}$$

Here, $\bar{\sigma}$ denotes the maximum singular value. From the definition of $\tilde{B}(s)$ and the Lyapunov equation (1.4.1), we obtain

$$\tilde{B}(j\omega)\tilde{B}^*(j\omega) = \Delta(j\omega)\Sigma_2 + \Sigma_2\Delta^*(j\omega). \tag{1.4.14}$$

The expression for $\tilde{C}^*(j\omega)\tilde{C}(j\omega)$ is obtained analogously [but using (1.4.2)] as

$$\tilde{C}^*(j\omega)\tilde{C}(j\omega) = \Sigma_2\Delta(j\omega) + \Delta^*(j\omega)\Sigma_2. \tag{1.4.15}$$

Substituting (1.4.14) and (1.4.15) into (1.4.13) with $\Sigma_2 = \sigma_n$ yields

$$\bar{\sigma}\big[G(j\omega) - G_r(j\omega)\big] = \sigma_n \lambda_{\max}^{1/2}\Big\{\big[1 + \Theta^{-1}(j\omega)\big]\big[1 + \Theta(j\omega)\big]\Big\}, \tag{1.4.16}$$

where $\Theta = (\Delta^{-1}(j\omega))^* \Delta(j\omega)$ is a scalar all-pass function. Hence $|\Theta(j\omega)| = 1$. Using the triangle inequality we obtain

$$\bar{\sigma}\,[G(j\omega) - G_r(j\omega)] \le \sigma_n\big[1 + |\Theta(j\omega)|\big] = 2\sigma_n. \tag{1.4.17}$$

Now define $G_k(s)$ as the truncated model of kth order, for which the Hankel singular values are $\sigma_1, \ldots, \sigma_k$. Let $E_k(s) = G_{k+1}(s) - G_k(s)$ for $k = 1, 2, \ldots, n-1$. Then $\bar{\sigma}[E_k(j\omega)] \le 2\sigma_{k+1}$ from (1.4.17). Noting that $G(s) - G_r(s) = \sum_{k=r}^{n-1} E_k(s)$, we have

$$\bar{\sigma}\,[G(j\omega) - G_r(j\omega)] \le \sum_{k=r}^{n-1} \bar{\sigma}\,[E_k(j\omega)] \le 2 \sum_{k=r+1}^{n} \sigma_k. \tag{1.4.18}$$

Conservatism of the error bound formula

The error bound formula is only a bound: the actual maximum error between the original model and the reduced order model may be smaller than the bound. It is then reasonable to ask how tight the bound is, and are there some circumstances when it may be very conservative. These questions were raised and some answers given by Enns (1984).

As an example that the bound can be tight, consider a transfer function

$$G(s) = \sum_{i=1}^{n} \frac{\beta_i}{s + \alpha_i}, \tag{1.4.19}$$

with $\alpha_i > 0$, $\beta_i > 0$. A realization for $G(s)$ is given by

$$\begin{aligned} A &= \operatorname{diag}[-\alpha_1, -\alpha_2, \ldots, -\alpha_n], \\ B &= \begin{bmatrix} \sqrt{\beta_1} & \sqrt{\beta_2} & \cdots & \sqrt{\beta_n} \end{bmatrix}^T = C^T. \end{aligned} \tag{1.4.20}$$

Equation (1.4.20) implies $P = Q$. Let V be orthogonal with $V^T P V = V^T Q V = \Sigma$ and set $\bar{A} = V^T A V$, $\bar{b} = \bar{c}^T = V^T b$. The associated grammians are both Σ. Now observe

$$G(0) = \bar{c}(-\bar{A})^{-1}\bar{b} = \operatorname{tr}\big[-\bar{A}^{-1}\bar{b}\bar{b}^T\big] = \operatorname{tr}\big[\bar{A}^{-1}(\Sigma\bar{A} + \bar{A}\Sigma)\big] = 2\operatorname{tr}\Sigma. \tag{1.4.21}$$

Obviously a similar calculation yields

$$G_r(0) = 2\operatorname{tr}\Sigma_1. \tag{1.4.22}$$

The error at $\omega = 0$ is

$$G(0) - G_r(0) = 2\,\mathrm{tr}\,\Sigma_2. \tag{1.4.23}$$

As a second example, consider a transfer function

$$G(s) = \sum_{r=1}^{n} \frac{\omega_i^2}{s^2 + 2\xi\omega_i s + \omega_i^2}, \tag{1.4.24}$$

where $\xi > 0$ and $\omega_1 < \omega_2 < \cdots < \omega_n$. When the ω_i are widely spaced and ξ is very small, the frequency response of G consists of a series of separated resonances.

One can in fact demonstrate that

$$\lim_{\xi \downarrow 0} G_r(s) = \lim_{\xi \downarrow 0} \sum_{i=1}^{r} \frac{\omega_i^2}{s^2 + 2\xi\omega_i s + \omega_i^2} \tag{1.4.25}$$

by using a similar approach to that of the first example.

Also,

$$\lim_{\xi \downarrow 0} \Sigma_2 = \frac{1}{2\xi} I_{2(n-r)}. \tag{1.4.26}$$

[The notation means that for very small positive ξ, the right hand side of (1.4.26) is a good approximation of Σ_2.] On the other hand,

$$\lim_{\xi \downarrow 0} \max_{\omega} \left| G(j\omega) - G_r(j\omega) \right| = \sup_{\omega} \left| \lim_{\xi \downarrow 0} \sum_{r+1}^{n} \frac{\omega_k^2}{\omega_k^2 - \omega^2 + 2\xi j\omega\omega_k} \right| = \lim_{\xi \downarrow 0} \frac{1}{2\xi}. \tag{1.4.27}$$

The normal error bound of $2\,\mathrm{tr}\,\Sigma_2$ then yields a quantity that is $2(n-r)$ times the actual bound. In this case, one might argue that the eigenvalues of Σ_2 are all the same, and so the unproved formula (based on discarding repeated Hankel singular values) would apply; then the bound and actual error become the same. However, arbitrarily small adjustments of the damping coefficient of the second order summands throws one back onto the $2\,\mathrm{tr}\,\Sigma_2$ formula, because Σ_2 will no longer be a multiple of the identity.

The error bound is thus a discontinuous function of $G(s)$ in the neighbourhood of a $G(s)$ which has repeated Hankel singular values. One might therefore expect it to be conservative in any situation where discarded singular values were very close.

L_1 error bound in impulse response

With $G(s)$ and $G_r(s)$ the full order and reduced order transfer functions respectively, we have concentrated on the error measure $\|G(j\omega) - G_r(j\omega)\|_\infty$. With $g(t)$ and $g_r(t)$ the impulse responses of $G(s)$ and $G_r(s)$, the L_1 measure

$$\int_0^\infty \left| g(t) - g_r(t) \right| dt \tag{1.4.28}$$

or

$$\int_0^\infty \bar{\sigma}\,[g(t) - g_r(t)]\,dt \tag{1.4.29}$$

can also be considered.

Some results, with fairly complicated derivation, can be found in Glover, Curtain and Partington (1988) and Lam and Anderson (1992). Here, we shall indicate a similar bound to those of the references, but obtained in a simpler way. This bound depends on the following result.

Lemma 1.4.1. *Let $G(s) = C(sI - A)^{-1}B$ be of degree n and asymptotically stable, with $\{A, B, C\}$ minimal. Let σ_i denote the Hankel singular values. Then, with $g(t)$ the associated impulse response,*

$$\|g\|_1 = \int_0^\infty \bar{\sigma}(g(t))dt \leq 2\sum_{i=1}^n \sigma_i. \tag{1.4.30}$$

Proof. The proof uses the fact that for any two matrices X, Y for which the product XY exists, $\bar{\sigma}(XY) \leq \bar{\sigma}(X)\bar{\sigma}(Y)$. Evidently,

$$\begin{aligned}
\|g\|_1 &= \int_0^\infty \bar{\sigma}\big[Ce^{At}B\big]dt \\
&= 2\int_0^\infty \bar{\sigma}\big[Ce^{As}e^{As}B\big]ds \quad (\text{Set } s = \tfrac{1}{2}t) \\
&\leq 2\int_0^\infty \bar{\sigma}\big[Ce^{As}\big]\bar{\sigma}\big[e^{As}B\big]ds \\
&= 2\int_0^\infty \lambda_{\max}^{1/2}\big[(\exp As)^T C^T C \exp As\big]\lambda_{\max}^{1/2}\big[(\exp As)BB^T(\exp As)^T\big]ds \\
&\leq 2\left[\int_0^\infty \lambda_{\max}\big[(\exp As)^T C^T C \exp As\big]ds\right]^{1/2} \\
&\quad\times\left[\int_0^\infty \lambda_{\max}\big[(\exp As)BB^T(\exp As)^T\big]ds\right]^{1/2} \\
&\leq 2\left\{\int_0^\infty \mathrm{tr}\big[(\exp As)^T C^T C \exp As\big]ds\right\}^{1/2} \\
&\quad\times\left\{\int_0^\infty \mathrm{tr}\big[(\exp As)BB^T(\exp As)^T\big]ds\right\}^{1/2} \\
&= 2\left\{\mathrm{tr}\int_0^\infty (\exp As)^T C^T C \exp As\, ds\right\}^{1/2} \\
&\quad\times\left\{\mathrm{tr}\int_0^\infty (\exp As)BB^T(\exp As)^T ds\right\}^{1/2} \\
&= 2(\mathrm{tr}\,Q\,\mathrm{tr}\,P)^{1/2},
\end{aligned} \tag{1.4.31}$$

where P, Q are the controllability and observability grammians. Choosing a balanced realization yields immediately

$$\|g\|_1 \leq 2\sum_{i=1}^{n} \sigma_i. \tag{1.4.32}$$

□

Now suppose that $G_r(s)$ is a degree r balanced truncation approximation of $G(s)$, which has degree n.

As we know

$$\left\|G(j\omega) - G_r(j\omega)\right\|_\infty \leq 2\sum_{r+1}^{n}{}' \sigma_i, \tag{1.4.33}$$

where $\sum'$ means perform the sum but disregard repeats of the same singular value. It is clear that $G - G_r$ has degree at most $n + r$; we show later the maximum singular value is overbounded by any L_∞ norm bound, *i.e.*,

$$\sigma_1(G - G_r) \leq 2\sum_{r+1}^{n}{}' \sigma_i. \tag{1.4.34}$$

Since $\sigma_i(G - G_r) \leq \sigma_1(G - G_r)$ for $i = 2, \ldots, n + r$, it follows from the Lemma that

$$\|g - g_r\|_1 \leq 4(n + r)\sum_{i=r+1}^{n}{}' \sigma_i. \tag{1.4.35}$$

Assume for convenience that the $\sigma_i(G)$ are distinct. An alternative bound comes from

$$\begin{aligned} \|g - g_r\|_1 &\leq \|g - g_{n-1}\|_1 + \|g_{n-1} - g_{n-2}\|_1 + \cdots + \|g_{r+1} - g_r\|_1 \\ &\leq 4\sum_{i=r+1}^{n}(2i - 1)\sigma_i. \end{aligned} \tag{1.4.36}$$

This may be a better bound. It can be improved when the Hankel singular values are not all distinct.

Error bounds for singular perturbation of a balanced realization

The fact that balanced truncation generally incurs an approximation error in the low-frequency region will be undesirable in some practical applications. An algorithm which produces zero error at zero frequency is obtained via an easy modification. The idea is to simply replace s with $1/s$ as follows, Green and Limebeer (1995):

1. Set $H(s) = G(1/s)$. If $G(s)$ has realization $\{A, B, C, D\}$ then $H(s)$ has realization $\{A^{-1}, A^{-1}B, CA^{-1}, D - CA^{-1}B\}$.

2. Let $H_r(s)$ be an rth order balanced truncation of $H(s)$. Note that $H_r(\infty)$ is chosen to equal $H(\infty)$

3. Set $G_r(s) = H_r(1/s)$. The algorithm leads to perfect steady-state performance since $G_r(0) = G(0)$ is equivalent to $H(\infty) = H_r(\infty)$.

More interestingly, if the initial realization of $G(s)$ is balanced, so is the realization of $H(s)$ and it turns out that, provided A_{22} in the balanced realization of $G(s)$ is nonsingular, $G_r(s)$ is exactly a singular perturbation of $G(s)$. (The calculations are not difficult.) Moreover, the controllability and observability grammians of G and H are the same (again the calculations are not difficult), and because the maximum frequency domain error in H is the same as the maximum frequency domain error in G, and the error in H is given by the formula applicable to balanced truncation, we have immediately that the error bound formula carries over to the approximation of G by singular perturbation of a balanced realization.

When the more general form of singular perturbation is used, resulting in matching of the transfer function matrix of the original and the reduced model at some positive ρ, again provided one starts with a balanced realization, the same error bound formula applies, see Liu and Anderson (1989) for details.

Situations where approximation will be difficult

It is helpful to note situations where reduction is likely to be difficult so that Σ will contain few diagonal entries which are very small. Suppose that $G(s)$ is strictly proper, has degree n and has $(n-1)$ unstable zeros. Then as ω runs from zero to ∞, the phase of $G(s)$ will change by $(2n-1)\pi/2$ radians. Much of this change may occur in the passband. Suppose $G_r(s)$ has degree $n-1$; it can have no more than $(n-2)$ zeros, since it is strictly proper. So, even if all zeros are unstable, the maximum phase shift when ω moves from 0 to ∞ is $(2n-3)\pi/2$. It follows that if $|G(j\omega)|$ remains large at frequencies when the phase shift has moved past $(2n-3)\pi/2$, approximation of $G(s)$ by $G_r(s)$ will necessarily be poor. The problem is at its most acute when $G(s)$ is an all-pass function, or the strictly proper part of an all-pass function. In all-pass functions, there are n stable poles and n unstable zeros. It is possible to show using the all-pass property that all Hankel singular values are 1; the error bound formula indicates for a reduction to order $n-1$ a maximum error of 2.

Avoiding numerical problems

Coordinate basis transformations which are not orthogonal are potential sources of numerical difficulty. Therefore, there is a risk in performing a model reduction using balanced truncation that the balancing step will be numerically hazardous. What can be done about this? Remember that the real end goal of a model reduction procedure is a transfer function matrix, or some state-variable realization of it which does not have to be balanced. In the light of this observation, it becomes reasonable to contemplate a reduction algorithm which starts with an arbitrary state variable realization of a transfer function matrix $G(s)$ and finds a state variable realization of a reduced order $G_r(s)$ without introducing non-orthogonal transformations, and with the property that had balanced truncation with perfect arithmetic been used on $G(s)$ the same $G_r(s)$ would have resulted.

Such algorithms are normal in commercial software implementations, (see *e.g.*, Matrix $_X$ Model Reduction Module, 1991—Integrated Systems Inc, Santa Clara, California.) We describe one such here, due to Safonov and Chiang (1989).

Let P and Q be the controllability and observability grammians of whatever state variable realization of $G(s)$ is initially available. (Of course, their computation may be demanding in terms of software.) Compute ordered Schur decompositions of the product PQ, with the eigenvalues of PQ in ascending order and descending order. Thus for some orthogonal matrices V_A and V_D there holds

$$V_A^T PQV_A = S_{\text{asc}}, \qquad V_D^T PQV_D = S_{\text{des}}. \tag{1.4.37}$$

Here S_{asc} and S_{des} are upper triangular. Now define submatrices as follows, with r the dimension of the reduced order system:

$$V_a = V_A \begin{bmatrix} 0 \\ I_r \end{bmatrix}, \qquad V_d = V_D \begin{bmatrix} I_r \\ 0 \end{bmatrix}, \tag{1.4.38}$$

and then form a singular value decomposition of $V_a^T V_d$:

$$U_L S U_R^T = V_a^T V_d. \tag{1.4.39}$$

Here U_L and U_R are orthogonal, while S is diagonal with positive entries. Now define transformation matrices

$$L = S^{-1/2} U_L^T V_a^T, \qquad R = V_d U_R S^{-1/2}. \tag{1.4.40}$$

Notice that $LR = I$. Finally define a state variable realization of $G_r(s)$ by

$$A_r = LAR, \qquad B_r = LB, \qquad C_r = CR \quad \text{and} \quad D_r = D. \tag{1.4.41}$$

Main points of the section

1. Balanced realization of a stable system give rise to controllability and observability grammians which are the same and diagonal.

2. Provided an inequality condition involving the diagonal elements of the grammians is satisfied, a truncation of a balanced realization is guaranteed to be stable and minimal.

3. An upper bound on the infinity norm of the frequency domain error is known for models obtained by balanced realization truncation. The value is equal to twice the sum of the truncated Hankel singular values.

4. Qualitative results are possible giving guidance as to when the upper bound formula is likely to be tight or not.

5. An upper bound is also available for the L_1 error in approximating an impulse response using balanced realization truncation.

1.5 Methods Involving Minimizing an Approximation Error Norm

The H_2 norm

The H_2 norm is widely used in control system synthesis. Consider the asymptotically stable system $G(s) = D + C(sI - A)^{-1}B$ and the reduced order stable model $G_r(s) = D_r + C_r(sI - A_r)^{-1}B_r$. Set $D_r = D$; then a realization of the error $E(s) = G(s) - G_r(s)$ is given by

$$\left[\begin{array}{cc} \tilde{A} & \tilde{B} \\ \tilde{C} & 0 \end{array}\right] = \left[\begin{array}{cc|c} A & 0 & B \\ 0 & A_r & B_r \\ \hline C & -C_r & 0 \end{array}\right]. \tag{1.5.1}$$

We will use the H_2 (or L_2) norm of $E(s)$ for evaluating the approximation error. The norm is calculated as follows:

$$\|E(s)\|_2 = \mathrm{tr}(\tilde{C}\tilde{P}\tilde{C}^T) = \mathrm{tr}(\tilde{B}^T\tilde{Q}\tilde{B}), \tag{1.5.2}$$

where $\tilde{P}$ and $\tilde{Q}$ are the solutions of the Lyapunov equations:

$$\tilde{A}\tilde{P} + \tilde{P}\tilde{A}^T + \tilde{B}\tilde{B}^T = 0, \qquad \tilde{Q}\tilde{A} + \tilde{A}^T\tilde{Q} + \tilde{C}^T\tilde{C} = 0. \tag{1.5.3}$$

Minimizing $\|E(s)\|_2$ with respect to A_r *etc.* leads to a nonconvex problem and there is no unique global minimum since the coordinate basis can be adjusted without changing the error; numerical algorithms aimed at solving the optimization problem can get stuck at local minima which are not global minima, see Meier and Luenberger (1967), Wilson (1970), Bernstein and Hyland (1985). One method to avoid the nonconvexity is to assign the poles of the reduced order model before minimizing and to regard the unknown parameters as the entries of C_r and D_r. The pre-assignment makes the problem simple so that solving one linear equation is enough to obtain the unique solution Kimura (1983). Of course, the achieved error may be considerably greater than that achievable in principle when poles can be varied.

Earlier, we have described model approximation by balanced truncation, and considered the question of the resulting frequency domain error. It is natural therefore to ask what can be said about the L_2 error when balanced truncation is used, and whether there is any relation between the frequency domain error and the L_2 error. In Kabamba

(1985) it is concluded that there is no simple relationship, and that balanced truncation may entail a very large L_2 error.

The argument goes as follows. Suppose that $\{A, B, C\}$ is a balanced triple, with Hankel singular values $\sigma_1, \sigma_2, \ldots, \sigma_n$ and $\sigma_i > \sigma_{i+1}$. Denote the ith row of B by b_i^T and the ith column of C by c_i. Then there exists ν_i with $\nu_i^2 = b_i^T b_i = c_i^T c_i = -(\sigma_i/2a_{ii})$. With P the controllability grammian, for any realization of a $G(s)$ there holds $\|G\|_2^2 = \operatorname{tr} CPC^T$ and so for the balanced realization $\|G\|_2^2 = \sum \sigma_i \nu_i^2$. Thus in a rough sense, the ith entry of the state vector contributes $(\sigma_i \nu_i^2)^{1/2}$ to the L_2 norm. It is clear that one could have $\sigma_n \ll \sigma_I$ and $\nu_n \gg \nu_i$ for $i \neq n$, and in particular for $i = n-1$. In this case order reduction by one can change the L_2 norm substantially. Also, by the triangle inequality,

$$\|G - G_{n-1}\|_2 \geq \|G\|_2 - \|G_{n-1}\|_2 .$$

So obviously the L_2 norm of the error can be substantial also.

H_∞ norm

There exists no good analytical method for minimizing the H_∞ norm of an additive error $E(s)$. We can of course try to apply a numerical method, probably iterative in character, to solve the problem starting with an initial reduced order model as a first iteration. However it does not seem possible to parametrize the unknown reduced order model so that the optimization problem is convex in the unknown parameters. Since Hankel norm approximation, which will be explained in the next section, gives a suboptimal but often not-far-from-optimal solution for the problem, it is at least possible to pick the initial reduced order model intelligently. Also, if one prescribes the poles, a convex problem can be obtained. Thus it is possible to choose as the prescribed poles those arising from a Hankel norm approximation problem and then use convex optimization.

Equation error techniques

Equation error techniques provide a device for turning an L_2 norm minimization problem into a quadratic minimization problem, for which there exists a linear equation defining the optimum. The idea is an old one in the identification literature. A brief sketch only of the technique will be given here for single-input, single-output systems. Consider the reduced order transfer function $G_r(s)$ with denominator $A(s)$ and numerator $B(s)$, given in the following form:

$$G_r(s) = \frac{B(s)}{A(s)} = \frac{b_q s^q + b_{q-1} s^{q-1} + \cdots + b_0}{a_r s^r + a_{r-1} s^{r-1} + \cdots + 1}.$$

If we form the additive error $E(s) = G(s) - G_r(s)$ and multiply it by $A(s)$, we obtain a new weighted error. Consider the following equation error:

$$\tilde{E}(s) = A(s)E(s) = A(s)G(s) - B(s). \tag{1.5.4}$$

Consider the response $y(t)$ of the transfer function $\tilde{E}(s)$ to a given input $h(t)$; then the L_2 norm of $y(t)$ can be calculated to evaluate the magnitude of the L_2 norm of $\tilde{E}(j\omega)$ weighted by $H(j\omega)$ as follows:

$$\|y\|_2^2 = \begin{bmatrix}1, a, b\end{bmatrix} W \begin{bmatrix}1 \\ a^T \\ b^T\end{bmatrix} = \begin{bmatrix}1, a, b\end{bmatrix} \begin{bmatrix} W_{11} & W_{12} & W_{13} \\ W_{12}^T & W_{22} & W_{23} \\ W_{13}^T & W_{23}^T & W_{33} \end{bmatrix} \begin{bmatrix}1 \\ a^T \\ b^T\end{bmatrix}, \tag{1.5.5}$$

where $a = [a_1, a_2, \ldots, a_r]$, $b = [b_0, b_1, \ldots, b_q]$ and where

$$W = \frac{1}{2\pi} \int_{-\infty}^{\infty} Z^*(j\omega) d\omega Z(j\omega) d\omega, \tag{1.5.6}$$

with the following definition:

$$Z(s) = \left[G(s), sG(s), \ldots, s^r G(s), -1, -s, \ldots, -s^q\right] H(s). \tag{1.5.7}$$

Here $H(s)$ is the Laplace transform of $h(t)$, with $H(s)$ rational and having d_H as the denominator order and n_H as the numerator order. If $G(s)$ and $H(s)$ are both asymptotically stable and if d_H and n_H satisfy the inequality $d_H > n_H + \max\{q, r + n_G - d_G\}$, then W is bounded so that the minimizing a and b satisfy the linear equation (necessary condition) see Sufleta (1984):

$$\begin{bmatrix} W_{22} & W_{23} \\ W_{23}^T & W_{33} \end{bmatrix} \begin{bmatrix} a^T \\ b^T \end{bmatrix} = - \begin{bmatrix} W_{12}^T \\ W_{13}^T \end{bmatrix}. \tag{1.5.8}$$

The reduced order model so obtained is dependent upon the choice of input $h(t)$, and the constraint on the orders of polynomials in $H(s)$ and $G_r(s)$ is restrictive. One can consider $H(s)$ as a prefilter for the linearized error: if $H(s)$ approximates $A(s)-1$, then the error measure comes close to the normal L_2 error measure. In discrete time systems, the method is more natural as there are no restrictions on the polynomial degrees and the weighting $H(s)$ can if desired be replaced by a unity weighting.

Equation error techniques can be applied to state space representations. Consider the state equation of a reduced order model:

$$\dot{x}_r = A_r x_r + B_r u. \tag{1.5.9}$$

Assume that $x(t)$ is the state vector of the original system and L is a constant $r \times n$ matrix with full rank. Define the error $e(t) = Lx(t) - x_r(t)$. Then it is straightforward to show that

$$\dot{e}(t) = A_r e(t) + \tilde{e}(t), \tag{1.5.10}$$

where

$$\tilde{e}(t) = (LA - A_r L)x(t) + (LB - B_r)u(t). \tag{1.5.11}$$

We now introduce the new index

$$J = \int_0^\infty \tilde{e}^T(t)\tilde{e}(t)dt. \tag{1.5.12}$$

This index can be viewed in two ways Obinata and Inooka (1976). First, $\tilde{e}(t)$ is the forcing term in error equation (1.5.10); thus minimizing it is a goal. Second, $\tilde{e}(t)$ can also be considered as an equation error resulting from substituting a linear combination of the original state Lx into the reduced order equation, *i.e.*,

$$\tilde{e}(t) = L\dot{x}(t) - \big(A_r Lx(t) + B_r u(t)\big). \tag{1.5.13}$$

The derivative is like the multiplication by s or a higher power of s in forming the index (1.5.5) using (1.5.6) and (1.5.7). The necessary conditions for minimizing J in (1.5.12) with respect to A_r and B_r are $\partial J/\partial A_r = 0$, $\partial J/\partial B_r = 0$, *i.e.*,

$$\begin{bmatrix} A_r & B_r \end{bmatrix} \begin{bmatrix} LW_x L^T & LW_{xu} \\ W_{xu}^T L^T & W_u \end{bmatrix} = L \begin{bmatrix} A & B \end{bmatrix} \begin{bmatrix} W_x L^T & W_{xu} \\ W_{xu}^T L^T & W_u \end{bmatrix}, \tag{1.5.14}$$

where

$$W_x = \int_0^\infty x(t)x^T(t)dt,\ W_{xu} = \int_0^\infty x(t)u^T(t)dt,\ W_u = \int_0^\infty u(t)u^T(t)dt. \tag{1.5.15}$$

It should be noted that W_x, W_{xu} and W_u are dependent upon the input $u(t)$ and the choice of state space coordinate basis. In the case of impulse or white noise inputs, the A_r and B_r which minimize J are given by

$$A_r = LAPL^T(LPL^T)^{-1}, \qquad B_r = LB, \tag{1.5.16}$$

where P is the controllability grammian of the original system. Because the error criterion is quadratic in the unknown parameters, the solution (1.5.16) to the minimization problem is unique and gives the global minimum. Further, under a white noise input the state covariance of (1.5.16) matches the covariance of Lx for the original system. If we select the state to output matrix C_r to be $CPL^T(LPL^T)^{-1}$, then the resulting reduced order model is a truncation after coordinate basis change of the original model. [See (1.2.10), and identify R with $PL^T(LPL^T)^{-1}$]. In the case of a step input, the closed form solution can also be given. It is not a truncated model but has the property of giving zero steady state error between the original system and the reduced order model under step input conditions, Obinata and Inooka (1983). It is easy to extend this method to the case where a pre-filter is used to provide a frequency dependent weighting of the equation error, Inooka and Obinata (1977). Equation error techniques are relevant in seeking matching of first order parameters such as Markov parameters or time moments and also of second order parameters or covariance parameters Mullis and Roberts (1976), Yousuff, Wagie and Skelton (1985), Obinata, Nakamura and Inooka (1988). This point will be developed in Section 1.7.

Main points of the section

1. There exists no good analytical method for minimizing the H_2 or H_∞ norm of the additive error in model reduction problems.

2. In balanced truncation approximation, a substantial H_2 error norm may result even with a small H_∞ error norm.

3. Reduced-order models which minimize the L_2 norm of an equation error can be easily calculated from linear algebraic equations.

1.6 Hankel Norm Approximation

The Hankel operator and the Hankel norm

The Hankel operator of a time-invariant linear system is an operator which maps past inputs to future outputs, formed from the impulse response of that system. Consider the minimal realization $\{A, B, C, D\}$ of a stable and proper system. Suppose inputs $u \in L_2(-\infty, 0]$ are applied up to time $t = 0$ and we examine outputs after time $t = 0$. We neglect any component of the output at $t = 0$ due to direct feedthrough of $u(0)$. These future outputs are determined by the convolution integral

$$y(t) = \int_{-\infty}^{0} Ce^{A(t-\tau)}Bu(\tau)d\tau, \qquad t > 0. \tag{1.6.1}$$

Equation (1.6.1) can be considered as a mapping $\Xi : L_2(-\infty, 0] \to L_2[0, \infty)$ and also as a mapping $\Gamma : L_2[0, \infty) \to L_2[0, \infty)$ (confirm this by setting τ for $-\tau$ in the right hand side of (1.6.1)). The latter mapping is taken as the definition of the Hankel operator:

$$(\Gamma u)(t) = \int_{0}^{\infty} Ce^{A(t+\tau)}Bu(\tau)d\tau = \int_{-\infty}^{0} Ce^{A(t-\tau)}Bu(\tau)d\tau. \tag{1.6.2}$$

Notice that the Hankel operator is defined so that it is independent of D. Although the Hankel operator is defined by Γ, it is often convenient to return to the equivalent operator Ξ, as it captures better the intuition associated with the causal mapping of $u(.)$ to $y(.)$. We shall do this without explicit comment below.

If $u \in L_2(-\infty, 0]$, then $y(t) = Ce^{At}x(0)$ for $t > 0$. Thus, we only need $x(0)$ to determine the future output. All (past) inputs which give rise to the same present $x(0)$ produce the same future output. Any two linearly independent initial ($t = 0$) states result in linearly independent future outputs because of the observability assumption. Hence the number of linearly independent outputs is n, the dimension of a minimal realization of the system. Therefore, the rank of the Hankel operator is equal to the McMillan degree of the system.

The induced norm of the system's Hankel operator is given by

$$\|G(s)\|_H = \sup_{u \in L^2[0,\infty)} \frac{\|\Gamma u\|_{L_2[0,\infty)}}{\|u\|_{L_2[0,\infty)}} = \sup_{u \in L^2(-\infty,0]} \frac{\|\Xi u\|_{L_2[0,\infty)}}{\|u\|_{L_2(-\infty,0]}}. \tag{1.6.3}$$

The future output energy "$\|\Xi u\|^2$" resulting from any past input u is at most the Hankel norm squared times the energy of the input, assuming the future input is zero. The Hankel norm is simply expressible using the controllability grammian P and observability grammian Q as follows:

$$\|G(s)\|_H = \lambda_{\max}^{1/2}(PQ). \tag{1.6.4}$$

(By some abuse of notation, we identify the Hankel norm of a transfer function matrix with the Hankel norm of its associated operator.) The claim is easily established. For the moment, fix $x(0)$. Then

$$\sup_{u \in L_2(-\infty,0)} \frac{\|\Xi u\|^2}{\|u\|^2} = \sup_{u \in L_2(-\infty,0)} \frac{x^T(0)Qx(0)}{\|u\|^2}. \tag{1.6.5}$$

It is well known that the minimum energy control taking $x(-\infty) = 0$ to $x(0)$ is $x^T(0)P^{-1}x(0)$. Hence for fixed $x(0)$,

$$\sup_{u \in L_2(-\infty,0)} \frac{\|\Xi u\|}{\|u\|} = \left[\frac{x^T(0)Qx(0)}{x^T(0)P^{-1}x(0)}\right]^{1/2}. \tag{1.6.6}$$

The maximum over all $x(0)$ is $\lambda_{\max}^{1/2}(QP)$ The Hankel singular values are, as defined in Section 1.4, the square roots of the eigenvalues of PQ, or the diagonal entries of P and Q in a balanced realization. The quantities are invariant under state space coordinate basis change. Evidently, the Hankel norm is the largest Hankel singular value. Below, we shall relate the other Hankel singular values to the Hankel operator.

The Hankel norm of a transfer function matrix is bounded above by the infinity norm of the same object, as seen in the following argument:

$$\begin{aligned}
\|G\|_H &= \sup_{u \in L^2(-\infty,0]} \frac{\|\Xi u\|_{L_2[0,\infty)}}{\|u\|_{L_2(-\infty,0]}} \\
&\le \sup_{u \in L^2(-\infty,0]} \frac{\|y\|_{L_2(-\infty,\infty)}}{\|u\|_{L_2(-\infty,0]}} \\
&\le \sup_{u \in L^2(-\infty,\infty)} \frac{\|y\|_{L_2(-\infty,\infty)}}{\|u\|_{L_2(-\infty,\infty)}} \\
&= \|G\|_\infty .
\end{aligned} \tag{1.6.7}$$

Moreover, the following relation holds.

$$\|G\|_H \le \|G - F\|_\infty , \tag{1.6.8}$$

where $F(s)$ is the transfer function matrix of any anticausal system. This can be established by a variant on the argument establishing that $\|G\|_H \leq \|G\|_\infty$. Suppose that $u \in L_2(-\infty, 0]$. Consider the response of $G - F$ to u. Over $(-\infty, 0)$, the response is due to G and F. However, over $(0, \infty)$, the response is due solely to G, since F is anticausal. Therefore, for such a u

$$\|(G-F)u\|^2_{L_2(-\infty,\infty)} \geq \|\Xi u\|^2_{L_2[0,\infty)} \,. \tag{1.6.9}$$

Hence

$$\begin{aligned} \|G\|^2_H &= \sup_{u \in L_2(-\infty,0]} \frac{\|\Xi u\|^2}{\|u\|^2_{L_2(-\infty,0]}} \leq \sup_{u \in L_2(-\infty,0]} \frac{\|(G-F)u\|^2_{L_2(-\infty,\infty)}}{\|u\|^2_{L_2(-\infty,0]}} \\ &\leq \sup_{u \in L_2(-\infty,\infty)} \frac{\|(G-F)u\|^2_{L_2(-\infty,\infty)}}{\|u\|^2_{L_2(-\infty,\infty)}} \\ &= \|G-F\|^2_\infty \,. \end{aligned} \tag{1.6.10}$$

Observe that if $F = 0$, the case $\|G\|_H \leq \|G\|_\infty$ results. As we shall see later, there exists a particular F such that the inequality (1.6.8) becomes an equality.

Schmidt decomposition

The Schmidt decomposition is the singular value decomposition of the Hankel operator Γ, which is given by

$$\Gamma(u) = \sum_{i=1}^{n} \sigma_i \langle u, v_i \rangle \, w_i \,, \tag{1.6.11}$$

where $v_i, w_i \in L_2[0, \infty)$, satisfying $\langle v_i, v_j \rangle = 0$ and $\langle w_i, w_j \rangle = 0$ for $i \neq j$ with $\|v_i\|_2 = 1$ and $\|w_i\|_2 = 1$ for all i. The singular values σ_i of the operator are what we have already called the Hankel singular values of the system $G(s)$, and we shall assume ordering with $\sigma_i \geq \sigma_{i+1}$. (The fact that the singular values as defined by (1.6.11) coincide with the square roots of the eigenvalues of PQ is established below.) The pair (v_i, w_i) is called a Schmidt pair. Assume that $\Gamma^\sim$is the adjoint of the Hankel operator Γ, so that $(\Gamma^\sim w)(t) = \int_0^\infty B^T e^{A^T(t+\tau)} C^T w(\tau) d\tau = \sum_{i=1}^n \sigma_i \langle w, u_i \rangle v_i$. (This is trivial to check.) Evidently, $\Gamma^\sim$is the Hankel operator of $G^T(s)$.

The following equations are a consequence of the orthogonality of v_i and w_i.

$$\Gamma v_i = \sigma_i w_i \,, \qquad \Gamma^\sim w_i = \sigma_i v_i \,. \tag{1.6.12}$$

Consider the $L_2[0, \infty)$ function $v(t) = B^T e^{A^T t} P^{-1} x_0$ with the controllability grammian P. Then

$$\begin{aligned} (\Gamma v)(t) &= C e^{At} \left(\int_0^\infty e^{A^T \tau} B B^T e^{A^T \tau} d\tau \right) P^{-1} x_0 \\ &= C e^{At} x_0. \end{aligned} \tag{1.6.13}$$

In the same way, if $w(t) = Ce^{At}x_0$, then

$$(\Gamma^{\sim} w)(t) = B^T e^{A^T t}\left(\int_0^{\infty} e^{A^T\tau}C^T Ce^{A\tau}d\tau\right)x_0 = B^T e^{A^T t}Qx_0, \tag{1.6.14}$$

where Q is the observability grammian.

If x_0 is chosen so that $Qx_0 = \sigma^2 P^{-1}x_0$, then $\Gamma^{\sim}\Gamma v = \sigma^2 v$ with $\|v\|_2^2 = x_0^T P^{-1}x_0 = x_0^T Qx_0/\sigma^2$ from (1.6.13) and (1.6.14). Therefore, the selection of a set of orthogonal vectors $x_i \in R^n$ such that $PQx_i = \sigma_i^2 x_i$, $x_i^T Qx_i = \sigma_i^2$ produces a set of Schmidt pairs:

$$v_i(t) = \sigma_i^{-2}B^T e^{A^T t}Qx_i, \qquad w_i(t) = \sigma_i^{-1}Ce^{At}x_i, \qquad i = 1, \ldots, n. \tag{1.6.15}$$

The above argument can of course be easily reversed.

Suboptimal Hankel norm approximation—a lower bound

We begin our discussion of approximation with a lower bound on the Hankel norm of the error in the case when the original system G of degree n is approximated by a stable system of degree r. Let $G_r \in RH_\infty$ have degree less than or equal to $r < n$, and let (v_i, w_i) for $i = 1, \ldots, n$ be the Schmidt pairs of G. Consider inputs

$$v = \sum_{i=1}^{r+1}\alpha_i v_i. \tag{1.6.16}$$

Since the v_i for $i = 1, \ldots, r+1$ span an $r+1$ dimensional space and the Hankel operator Γ_r of G_r has rank $\leq r$, we can select the α_i, which are not all zero, such that $(\Gamma_r v)(t) = 0$. Using such a v, we obtain

$$\begin{aligned}\|(\Gamma - \Gamma_r)v\|_2^2 = \|\Gamma v\|_2^2 = \left\|\sum_{i=1}^{r+1}\alpha_i\sigma_i w_i\right\|_2^2 &= \sum_{i=1}^{r+1}\alpha_i^2\sigma_i^2 \\ &\geq \sigma_{r+1}^2\sum_{i=1}^{r+1}\alpha_i^2 = \sigma_{r+1}^2\|v\|_2^2.\end{aligned} \tag{1.6.17}$$

This establishes that

$$\|G - G_r\|_H \geq \sigma_{r+1}. \tag{1.6.18}$$

If equality holds in (1.6.18), then it obviously holds in (1.6.17). If $\sigma_r > \sigma_{r+1}$, then $\sigma_i > \sigma_{r+1}$ for $i = 1, \ldots, r$ and we conclude that $\alpha_i = 0$ for $i = 1, \ldots, r$. Since the α_i are not all zero, α_{r+1} cannot be zero and we must have $\Gamma_r v_{r+1} = 0$. This argument obviously yields $\Gamma_r v_j = 0$ for any $j > r+1$ such that $\sigma_j = \sigma_{r+1}$; one simply replaces the last summand in (1.6.16) by $\alpha_j v_j$. We shall later show that there do exist G_r achieving equality in (1.6.18).

Since the infinity norm is never smaller than the Hankel norm, (1.6.18) means also

$$\|G - G_r\|_\infty \geq \sigma_{r+1}(G). \tag{1.6.19}$$

Some system theory: counting unstable eigenvalues

In this subsection, we review a generalization of the Lemma of Lyapunov, which allows us to count the number of right half plane and left half plane eigenvalues of a square matrix.

Lemma 1.6.1. *Let A and $P = P^T$ be $n \times n$ matrices satisfying*

$$PA^T + AP + BB^T = 0, \tag{1.6.20}$$

for some P. Suppose that P is nonsingular and A has no purely imaginary eigenvalues. Then the number of eigenvalues of A with negative (positive) real parts equals the number of positive (negative) eigenvalues of P

Proof. Let $\epsilon > 0$ be sufficiently small that $A - \epsilon P$ has the same number of eigenvalues in $\mathrm{Re}[s] > 0$ and $\mathrm{Re}[s] < 0$ as A. Now

$$P\left(A^T - \epsilon P\right) + \left(A - \epsilon P\right)P + \left(BB^T + 2\epsilon P^2\right) = 0 \tag{1.6.21}$$

and $BB^T + \epsilon P^2$ is a positive definite matrix. Then the result follows by Theorem 2.4.10 of Horn and Johnson (1991). □

Suboptimal Hankel norm approximation

As a first step towards showing that there exist G_r achieving equality in (1.6.18), we will demonstrate a construction of a stable $G_r(s)$ of degree r and an antistable (all poles in the open right half plane) $F(s)$ of degree no greater than $n - r$ such that $\|G - G_r - F\|_\infty \leq \gamma$ for any value of γ in the open interval $(\sigma_{r+1}(G), \sigma_r(G))$. [We cope with multiple singular values as follows. Suppose $\sigma_{i-1} > \sigma_i = \sigma_{i+1} = \cdots = \sigma_{i+j} > \sigma_{i+j+1}$. Then we demonstrate the construction of a stable G_i with $\|G - G_i - F\|_\infty \leq \gamma$ for any value of γ in the interval $(\sigma_{i+j+1}, \sigma_i)$]. The result shows that the lower bound above is an infimum, but does not show that it is attainable, which is harder. We can regard $G_r(s)$ as a suboptimal Hankel norm approximation, suboptimal because it does not attain the infimum.

For constructing Hankel norm approximations, we need a tool to embed the given system $G(s)$ into an all-pass system. The tool is called all-pass embedding. Assume that $G \in RH_\infty$. Assume also that $G(s)$ is $p \times m$ with minimal realization $\{A, B, C, D\}$. Consider the augmented system:

$$G_a = \begin{bmatrix} G & 0 \\ 0 & 0 \end{bmatrix} = \begin{bmatrix} A & B & 0 \\ C & D & 0 \\ 0 & 0 & 0 \end{bmatrix} = \begin{bmatrix} A & B_a \\ C_a & D_a \end{bmatrix}. \tag{1.6.22}$$

where the additional zero blocks are chosen so that $G_a(s)$ has $(m + p)$ inputs and $(p + m)$ outputs, and so is square. We are going to show that there exists Q_a with realization

$$Q_a = \left[\begin{array}{c|c} \hat{A} & \hat{B} \\ \hline \hat{C} & \hat{D} \end{array}\right],$$

such that the error system

$$E_a = G_a - Q_a \tag{1.6.23}$$

satisfies

$$E_{a*}E_a = E_a E_{a*} = \gamma^2 I, \tag{1.6.24}$$

under the proviso that $\gamma > \sigma_{r+1}(G)$ and $\gamma < \sigma_r(G)$. Here, E_{a*} denotes $E_a^T(-s)$, and equation (1.6.24) means that the transfer function $E_a(s)$ is a scaled all-pass. We will also show that Q_a has a stable part of degree r. When $G_r + F$ is formed from Q_a through deletion of the last p columns and m rows, with G_r stable of degree r and F unstable, from (1.6.23) and (1.6.24) we will have $\|G - G_r\|_H \leq \|G - G_r - F\|_\infty \leq \gamma$. In the square G case, there holds $\|G - G_r - F\|_\infty = \gamma$.

With P the controllability grammian and Q the observability grammian of $G(s)$ and with $E = QP - \gamma^2 I$, define

$$\hat{D} = \begin{bmatrix} D & \gamma I_p \\ \gamma I_m & 0 \end{bmatrix}, \tag{1.6.25}$$

$$\hat{C} = \left(D_a - \hat{D}\right) B_a^T + C_a P, \tag{1.6.26}$$

$$\hat{B} = E^{-1}\left(Q B_a + C_a^T\left(D_a - \hat{D}\right)\right), \tag{1.6.27}$$

$$\hat{A} = -A^T - \hat{B}B_a^T = -E^{-1}\left(A^T E + C_a^T \hat{C}\right). \tag{1.6.28}$$

The transfer function matrix $E_a(s)$ is given by

$$E_a = \left[\begin{array}{cc|c} A & 0 & B_a \\ 0 & \hat{A} & \hat{B} \\ \hline C_a & -\hat{C} & D_a - \hat{D} \end{array}\right] = \left[\begin{array}{c|c} A_e & B_e \\ \hline C_e & D_e \end{array}\right]. \tag{1.6.29}$$

It is easy to confirm that the following Lyapunov-like equation holds:

$$P_e A_e^T + A_e P_e + B_e B_e^T = 0, \tag{1.6.30}$$

where P_e is given by

$$P_e = \begin{bmatrix} P & I \\ I & E^{-1}Q \end{bmatrix} \tag{1.6.31}$$

and is guaranteed to be nonsingular. It is also easy to verify that

$$P_e C_e^T + B_e D_e^T = 0. \tag{1.6.32}$$

The fact that $D_e^T D_e = \gamma^2 I$ together with (1.6.30) through (1.6.32) yields $E_{a*} E_a = \gamma^2 I$. Next, let us argue that $\hat{A}$ in the realization of $Q_a(s)$ has no purely imaginary axis eigenvalues. To obtain a contradiction, suppose $\hat{A}^T x = \lambda x$ for some $\text{Re}(\lambda) = 0$ and $x \neq 0$. Consider the 2-2 block entry of (1.6.30), which is

$$E^{-1} Q \hat{A}^T + \hat{A} E^{-1} Q + \hat{B} \hat{B}^T = 0. \tag{1.6.33}$$

Pre-multiplying and post-multiplying by x^* and x yields $\hat{B}^T x = 0$. Now the 1-2 block entry of (1.6.30) yields

$$\hat{A}^T + A + B_a \hat{B}^T = 0 \tag{1.6.34}$$

and so $Ax = -\lambda x$, which contradicts the fact that A is (asymptotically) stable.

Since both A and $\hat{A}$ have no imaginary axis eigenvalues, neither does A_e. In the light of the Lemma of the previous subsection, the nonsingularity of P_e and the absence of imaginary axis eigenvalues of A_e we can count the number of left half plane and right half plane eigenvalues of A_e. The matrix P is easily seen to have the same inertia as the matrix

$$\begin{bmatrix} P & 0 \\ 0 & E^{-1}Q - P^{-1} \end{bmatrix} = \begin{bmatrix} P & 0 \\ 0 & \gamma^2 (PE)^{-1} \end{bmatrix} \tag{1.6.35}$$

and the choice of γ ensures that $PE = P(QP - \gamma^2 I)$ has r positive and $n-r$ negative eigenvalues. It is then immediate that $\hat{A}$ has r negative real part eigenvalues and $n-r$ positive real part eigenvalues. Let $\hat{B}$ be partitioned as $[\hat{B}_1 \ \ \hat{B}_2]$ in the same manner as B_a. Notice that when $G_r + F$ is formed from Q_a, through deletion of the last p columns and m rows, $\hat{B}$ will be replaced by $\hat{B}_1$ and similarly for $\hat{C}$ and $\hat{D}$. We shall now argue that the left half plane eigenvalues of $\hat{A}$, which will be poles of G_r, are all controllable from $\hat{B}_1$. A corresponding observability conclusion follows in the same way.

To obtain a contradiction, suppose that $\hat{A}^T x = \lambda x$ and $\hat{B}_1^T x = 0$ for some $\text{Re}[\lambda] < 0$ and $x \neq 0$. Now the 1-2 block entry of (1.6.30) is

$$0 = \hat{A}^T + A + B_a \hat{B}^T = \hat{A}^T + A + B \hat{B}_1^T. \tag{1.6.36}$$

It follows that $Ax = -\lambda x$, which contradicts the asymptotic stability of A.

In general, it cannot be shown that the right half plane eigenvalues of $\hat{A}$ are all controllable. It follows that $G_r(s)$ has degree r while the unstable part of $Q_a(s)$ after truncation of the last m rows and p columns, namely $F(s)$, has degree at most $n-r$ and may have degree less than $n-r$.

When γ exceeds σ_1, $E^{-1}Q$ is negative definite and the 2-2 block of (1.6.30) and the fact that $\hat{A}$ has no imaginary axis eigenvalue imply that $(\hat{A}, \hat{B})$ is controllable. Similarly, $(\hat{A}, \hat{C})$ is observable. Then $F(s)$ has degree equal to $n-1$.

Linear fractional transformations

In the previous subsection, we have described how one particular suboptimal Hankel norm approximation can be found. Our aim now is to describe how all suboptimal approximations can be made. For this purpose, we need to use a device termed a Linear Fractional Transformation, and in particular appeal to a number of properties of such transformations. This subsection is devoted towards summarizing those properties, and the next subsection uses them in describing the class of all suboptimal Hankel norm approximants.

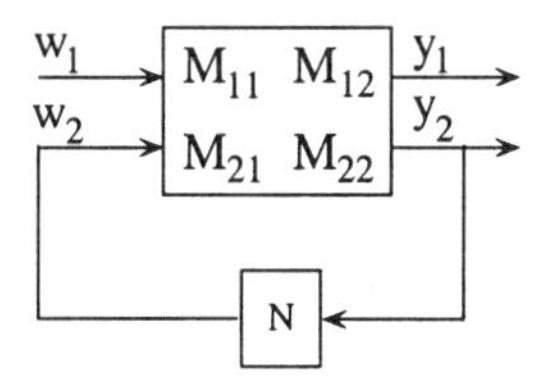

Figure 1.6.1. A lower linear fractional transformation

Consider the arrangement of Figure 1.6.1. The $M_{ij}(s)$ and $N(s)$ are all proper transfer function matrices of compatible dimensions and in order that the loop be well-defined, there holds

$$\det[I - M_{22}N](\infty) \neq 0. \tag{1.6.37}$$

The closed-loop transfer function from w_1 to y_1 is

$$G = M_{11} + M_{12}N\,(I - M_{22}N)^{-1}\,M_{12}. \tag{1.6.38}$$

This is termed a lower fractional transformation (LFT), and one uses the notation

$$G = \mathcal{F}_l\,[M, N]\,. \tag{1.6.39}$$

(The subscript l stands for "lower". Clearly, we can define an upper LFT also.)

An alternative view is provided from network theory, see Figure 1.6.2. The matrix M is the scattering matrix of a network, and a second network of scattering matrix N is used to terminate the first network. The scattering matrix of the combination is then G.

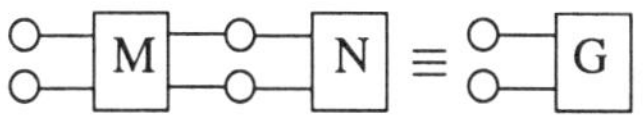

M N ≡ G

Figure 1.6.2. Network view of LFT

In Green and Limebeer (1995) a number of properties of LFTs are established. These are of several kinds: properties dealing with the magnitude or norm of G, M

and N, an invertibility property and properties dealing with the poles (given conditions on magnitudes).

We sum up the magnitude results in a theorem, see Section 4.3.2 of Green and Limebeer (1995). A number of the results will be no surprise, reflecting properties such as the fact that an interconnection of passive or lossless networks is again passive or lossless.

Theorem 1.6.2. *Consider the LFT $G = \mathcal{F}_l(M, N)$ under the well-posedness condition* (1.6.37). *Then*

1. $\|M\|_\infty \leq 1$ *and* $\|N\|_\infty \leq 1$ *imply* $\|G\|_\infty \leq 1$
2. $M_* M = I$ *and* $N_* N = I$ *imply* $G_* G = I$
3. $M_* M = I$ *and* $\|G\|_\infty < 1$ *implies* $M_{21}(j\omega)$ *has full column rank for all real* ω
4. *Suppose* $M_* M = I$ *and* $M_{21}(j\omega)$ *has full row rank for all real* ω. *Then*
   (a) $\|G\|_\infty > 1$ *if and only if* $\|N\|_\infty > 1$
   (b) $G_* G = I$ *if and only if* $N_* N = I$
   (c) $\|N\|_\infty \leq 1$ *if and only if* $\|G\|_\infty \leq 1$
   (d) $\|N\|_\infty < 1$ *and* M_{21} *is square if and only if* $\|G\|_\infty < 1$

The invertibility question is one of guaranteeing that, for a given M, an N can be found providing a prescribed G. The answer is as follows.

Theorem 1.6.3. *Let transfer function matrices M, G be prescribed, with $\|M\|_\infty$ and $\|G\|_\infty$ finite. A sufficient condition for N to exist such that $G = \mathcal{F}_l(M, N)$ and $\|N\|_\infty$ is finite (given dimension compatibility) is that $M_{12}(j\omega)$, $M_{21}(j\omega)$ are square and nonsingular for all real ω, and $M_{22}(\infty) = 0$.*

This may be proven by "solving" (1.6.38) for N.

Finally, we have the following result, Lemma 4.3.4 of Green and Limebeer (1995). In the statement, the words $X(s)$ has α poles in the region B for a rational proper $X(s)$ should be taken to mean that the 'A' matrix in a minimal state-variable realization has α eigenvalues in the region B.

Theorem 1.6.4. *Suppose that*

$$M = \left[\begin{array}{c|cc} A & B_1 & B_2 \\ \hline C_1 & 0 & D_{12} \\ C_2 & D_{21} & 0 \end{array}\right], \tag{1.6.40}$$

where $D_{12} = M_{12}(\infty)$ and $D_{21} = M_{21}(\infty)$ are nonsingular. Suppose that A has exactly r eigenvalues in $\mathrm{Re}[s] < 0$, *and $A - B_1 D_{12}^{-1} C_2$ and $A - B_2 D_{21}^{-1} C_1$ have all eigenvalues in* $\mathrm{Re}[s] > 0$. *Suppose that $\|M_{22} N\|_\infty < 1$ for some square rational $N(s)$ of appropriate dimensions. Then $G = \mathcal{F}_l(M, N)$ has exactly $r + l$ poles in* $\mathrm{Re}[s] < 0$ *if and only if N has exactly l poles in* $\mathrm{Re}[s] < 0$.

We remark that the condition $\|M_{22}N\|_\infty < 1$ is a type of small gain condition. Roughly speaking, it prevents changes from the pole distribution of M, N in open loop as a result of closing the loop. In open loop of course, M and N have exactly $(r+l)$ stable poles.

The class of all suboptimal Hankel norm approximants

In discussing suboptimal Hankel norm approximation, we started with a $G(s)$, added zero rows and columns to it to make $G_a(s)$, and then found a square $Q_a(s)$, with realization $\{\hat{A}, \hat{B}, \hat{C}, \hat{D}\}$ such that $G_a - Q_a = E_a$, a scaled all-pass satisfying $E_{a*}E_a = \gamma^2 I$. Here, $\gamma \in (\sigma_{r+1}(G), \sigma_r(G))$.

The determination of all approximants is now straightforward. Let $RH_\infty^-(r)$ denote the set of real rational transfer function matrices such that the A matrix of an associated minimal state variable realization has r eigenvalues in $\text{Re}[s] < 0$. We claim that the set of all $Q(s) \in RH_\infty^-(r)$ that satisfy

$$\|G(s) - Q(s)\|_\infty < \gamma \tag{1.6.41}$$

is given by

$$Q(s) = \mathcal{F}_l(Q_a, U) \qquad U \in RH_\infty^- \qquad \|U\|_\infty < \frac{1}{\gamma}. \tag{1.6.42}$$

Of course, RH_∞^- is the set of real rational transfer functions with all poles in $\text{Re}[s] > 0$. This may be proved as follows. Suppose that $Q(s)$ satisfies (1.6.41) and recall that

$$Q_a = \left[\begin{array}{c|cc} \hat{A} & \hat{B}_1 & \hat{B}_2 \\ \hline \hat{C}_1 & D & \gamma I_p \\ \hat{C}_2 & \gamma I_m & 0 \end{array}\right].$$

Minor manipulation of the earlier expressions for $\hat{A}$ *etc.* yields the following two equations:

$$\begin{bmatrix} \hat{A} - \lambda I & \hat{B}_1 \\ \hat{C}_2 & \gamma I \end{bmatrix} = \begin{bmatrix} -A^T - \lambda I & \hat{B}_1 \\ 0 & \gamma I \end{bmatrix} \begin{bmatrix} I & 0 \\ -B^T & I \end{bmatrix} \tag{1.6.43}$$

and

$$\begin{bmatrix} \hat{A} - \lambda I & \hat{B}_2 \\ \hat{C}_1 & \gamma I \end{bmatrix} = \begin{bmatrix} I & -E^{-1}C^T \\ 0 & I \end{bmatrix} \begin{bmatrix} -E^{-1}A^T E - \lambda I & 0 \\ \hat{C}_1 & \gamma I \end{bmatrix}, \tag{1.6.44}$$

from which it is clear that Q_{a12} and Q_{a21} are nonsingular on the imaginary axis and have all zeros in $\text{Re}[s] > 0$, while also $Q_{a22}(\infty) = 0$. By Theorem 1.6.3 of the previous subsection, (1.6.42) may be solved for U with $\|U\|_\infty$ finite. Now observe that the zero blocks in G_a yield

$$G - Q = \mathcal{F}_l(G_a - Q_a, U) = \mathcal{F}_l(E_a, U) \tag{1.6.45}$$

or

$$\gamma^{-1}(G - Q) = \mathcal{F}_l\left(\gamma^{-1}E_a, \gamma U\right). \tag{1.6.46}$$

Since $\gamma^{-1}E_a$ is all-pass and (1.6.41) holds, by Part (d)4 of Theorem 1.6.2, $\|\gamma U\|_\infty < 1$.

Last, consider $Q = \mathcal{F}_l(Q_a, U)$ in the light of Theorem 1.6.4. The zeros of Q_{a12}, Q_{a21} are in Re$[s] > 0$, and Q_a and Q both have precisely r poles in Re$[s] < 0$. The zero blocks in G_a mean that

$$[Q_a]_{22} = [Q_a - G_a]_{22} = [-E_a]_{22} \tag{1.6.47}$$

and so

$$\left\| Q_{a22} \right\|_\infty \leq \gamma \tag{1.6.48}$$

and

$$\left\| Q_{a22}U \right\|_\infty \leq \left\| Q_{a22} \right\|_\infty \left\| U \right\|_\infty < 1. \tag{1.6.49}$$

Hence $U \in RH_\infty^-$ by Theorem 1.6.4.

Obtaining some optimal Hankel norm approximations of a square transfer function matrix

In earlier subsections, we have obtained all reduced order models which satisfy a Hankel norm error bound exceeding the infimum attainable. The problem of optimal (as opposed to suboptimal) Hankel norm approximation is to find $G_r(s)$ of McMillan degree at most r which achieves equality in (1.6.18), *i.e.*, yields a Hankel norm γ for the error:

$$\gamma = \sigma_{r+1}(G). \tag{1.6.50}$$

The construction given earlier for the suboptimal approximation problem cannot be used here, as it involves the inverse of a matrix E which in the optimal case is clearly singular. The broad approach is however similar.

We shall first consider square $G(s)$, and find a limited family of optimal approximations. Then in later subsections we shall remove the squareness restriction, and expand modestly the family of optimal approximations. Finally, we will consider the construction of all optimal approximations.

Suppose that the realization of G, assumed for the moment to be square, is chosen so that the grammians are of the form

$$P = \begin{bmatrix} P_1 & 0 \\ 0 & \sigma_{r+1} I_l \end{bmatrix}, \qquad Q = \begin{bmatrix} Q_1 & 0 \\ 0 & \sigma_{r+1} I_l \end{bmatrix}, \tag{1.6.51}$$

with l the multiplicity of σ_{r+1}, which is strictly less than σ_r. Partition the realization of G conformably with P and Q to obtain

$$A = \begin{bmatrix} A_{11} & A_{12} \\ A_{21} & A_{22} \end{bmatrix}, \qquad B = \begin{bmatrix} B_1 \\ B_2 \end{bmatrix}, \qquad C = \begin{bmatrix} C_1 & C_2 \end{bmatrix}. \tag{1.6.52}$$

We prove below that there exists an orthogonal matrix U satisfying

$$B_2 = -C_2^T U. \tag{1.6.53}$$

Define

$$E_1 = Q_1 P_1 - \sigma_{r+1}^2 I \tag{1.6.54}$$

and define a realization $\{\hat{A}, \hat{B}, \hat{C}, \hat{D}\}$ of $Q(s)$ by

$$\hat{A} = E_1^{-1}(\sigma_{r+1}^2 A_{11}^T + Q_1 A_{11} P_1 - \sigma_{r+1} C_1^T U B_1^T), \tag{1.6.55}$$

$$\hat{B} = E_1^{-1}(Q_1 B_1 + \sigma_{r+1} C_1^T U), \tag{1.6.56}$$

$$\hat{C} = C_1 P_1 + \sigma_{r+1} U B_1^T, \tag{1.6.57}$$

$$\hat{D} = D - \sigma_{r+1} U. \tag{1.6.58}$$

We define the error system $E(s) = G(s) - Q(s)$ with realization $\{A_e, B_e, C_e, D_e\}$ constructed from $\{A, B, C, D\}$ and $\{\hat{A}, \hat{B}, \hat{C}, \hat{D}\}$ as in the suboptimal case.

To check that U exists, observe that the controllability and observability grammian equations for $G(s)$ imply

$$\sigma_{r+1}^2 (A_{22} + A_{22}^T) + B_2 B_2^T = 0, \tag{1.6.59}$$

$$\sigma_{r+1}^2 (A_{22}^T + A_{22}) + C_2^T C_2 = 0, \tag{1.6.60}$$

from which $B_2 B_2^T = C_2^T C_2$. Because $G(s)$ is square, B_2 and C_2^T have the same number of columns and the existence claim for U is then immediate. Now mimicking the earlier calculations, we can verify that

$$P_e = \begin{bmatrix} P_1 & 0 & I \\ 0 & \sigma_{r+1} I_l & 0 \\ I & 0 & E_1^{-1} Q_1 \end{bmatrix} \tag{1.6.61}$$

satisfies $P_e A_e^T + A_e P_e + B_e B_e^T = 0$ and $P_e C_e^T + B_e D_e^T = 0$ and then that $E_*(s)E(s) = \sigma_{r+1}^2 I$. The argument that $\hat{A}$ has no purely imaginary axis eigenvalue, and then that all left half plane eigenvalues of $\hat{A}$ are controllable from the first m columns of $\hat{B}$ and observable from the first $p = m$ rows of $\hat{C}$ goes as before.

One can check in a similar manner to previously that P_e has $n+r$ positive eigenvalues and $n-r-l$ negative eigenvalues, and $\hat{A}$ has r negative real part and $(n-r-l)$ positive real part eigenvalues. It follows that we have constructed $G_r(s)$ with exactly r stable poles and $F(s)$ with up to $(n-r-l)$ unstable poles—a fewer number than for the suboptimal case—such that

$$G(s) - G_r(s) - F(s) = E(s) \tag{1.6.62}$$

is a scaled all-pass, *i.e.*,

$$\|G(s) - G_r(s) - F(s)\|_\infty = \sigma_{r+1}. \tag{1.6.63}$$

Since for any G_r with exactly r stable poles, there holds $\sigma_{r+1} \leq \|G - G_r\|_H$ (the lower bound established earlier), and since $\|G - G_r\|_H \leq \|G - G_r - F\|_\infty$ for any F with entirely unstable poles (see equation (1.6.5)) it follows that

$$\|G(s) - G_r(s)\|_H = \sigma_{r+1}. \tag{1.6.64}$$

There are two cases where we can be definitive about the degree of $F(s)$. First, if $r = n - l$, then $F(s)$ has degree zero. [Observe that $n - r - l$ cannot be positive.] Second, if $r = 0$, then (arguing as for the suboptimal case) $F(s)$ has degree equal to n minus the multiplicity of the largest Hankel singular value of $G(s)$.

Above, we have explained how a particular square stable degree r optimal Hankel norm approximation $G_r(s)$ of a square degree n stable $G(s)$ can be constructed with the aid of a scaled all-pass $E(s)$ for which $E = G - G_r - F$ and $F(s)$ has all poles in the right half plane. It is also true that ANY optimal G_r can be associated with an unstable F such that the resulting E is necessarily a scaled all-pass. To see this, suppose that an optimal G_r is available, so that $\sigma_1(G - G_r) = \sigma_{r+1}(G)$. Applying the result we have proved to $G - G_r$ shows the existence of an unstable F such that $G - G_r - F$ is a scaled all-pass, as required. Note though that the degree of this F can only be bounded by the degree of $G - G_r$ minus the multiplicity of the maximum Hankel singular value of $G - G_r$; thus F may have degree $n + r - 1$. This observation of course falls short of providing a parametrization of all solutions of the optimal approximation problem; we shall do this below.

Obtaining some optimal Hankel norm approximations of a nonsquare transfer function matrix

Suppose now that $G(s)$ is $p \times m$ with $p \neq m$. Augment it as was done in the suboptimal approximation problem by adding p columns of zeros and m rows of zeros, thereby providing a square $G_a(s)$. Proceed for G_a in like manner to the square G case, but with a twist on the choice of U. Let $\bar{U}$ be any matrix satisfying

$$B_{a2} + C_{a2}^T \bar{U} = 0, \qquad \bar{U}^* \bar{U} = I, \tag{1.6.65}$$

so that there holds

$$\begin{bmatrix} B_2 & 0 \end{bmatrix} + \begin{bmatrix} C_2^T & 0 \end{bmatrix} \bar{U} = 0. \tag{1.6.66}$$

Let U be the 1-1 block of $\bar{U}$. Then

$$U^T U \leq I \tag{1.6.67}$$

and

$$B_2 = -C_2^T U. \tag{1.6.68}$$

It will turn out below that the formulas for the optimal approximant involve U but not the other parts of $\bar{U}$. In fact then, we can forget if we wish about $\bar{U}$ and just work with any U satisfying the last two equations above, for such a U could always be embedded in a $\bar{U}$ satisfying (1.6.65).

As for the case of square optimal approximation, we construct $Q_a(s)$ with realization $\{\hat{A}, \hat{B}, \hat{C}, \hat{D}\}$ such that

$$G_a - Q_a = E_a, \tag{1.6.69}$$

with $E_{a*} E_a = \sigma_{r+1}^2 I$. With G_{ar} the causal part of Q_a,

$$\| G_a - G_{ar} \|_H = \sigma_{r+1}. \tag{1.6.70}$$

Since $G - G_r$ is a submatrix of $G_a - G_{ar}$, there holds

$$\| G - G_r \|_H \leq \| G_a - G_{ar} \|_H = \sigma_{r+1}. \tag{1.6.71}$$

On the other hand, σ_{r+1} is a lower bound on the Hankel norm of the error of any approximation of G by a G_r with exactly r stable poles. Hence

$$\| G - G_r \|_H = \sigma_{r+1}. \tag{1.6.72}$$

In summary, we can use the formulas for the square case for generating $\hat{A}$, $\hat{B}$, $\hat{C}$, and $\hat{D}$ but with $\bar{U}$ replacing U, B_a replacing B *etc.* From the formulas, we derive a stable G_{ar} and an antistable F_a with $G_a - G_{ar} - F_a$ a scaled all-pass E_a. On deleting the last p columns and m rows, G_r and F are obtained.

Further, the submatrix argument gives

$$\| G - G_r - F \|_\infty \leq \sigma_{r+1} \tag{1.6.73}$$

and optimality again gives

$$\|G - G_r - F\|_\infty = \sigma_{r+1}. \tag{1.6.74}$$

Review of the formulas will show that the occurrences of $\bar{U}$ in the formulas for $\hat{A}$, $\hat{B}$ and $\hat{C}$ all involve a product of the type $\bar{U}B_{a1}^T$ or $C_{a1}^T\bar{U}$, and taking account of the introduced zeros, these products simplify as follows:

$$C_{a1}^T\bar{U}B_{a1}^T = C_1^TUB_1^T, \tag{1.6.75}$$

$$C_{a1}^T\bar{U} = \begin{bmatrix} C_1^TU & 0 \end{bmatrix}, \tag{1.6.76}$$

$$\bar{U}B_{a1}^T = \begin{bmatrix} UB_1^T \\ 0 \end{bmatrix}. \tag{1.6.77}$$

Also, when the last p columns and the last m rows are deleted from $\hat{D}$ in forming $G_r(s) + F(s)$, the matrix $D - \sigma_{r+1}U$ will be left. In the final analysis then, $\bar{U}$ will disappear from the formulas for the state variable realization of $G_r(s) + F(s)$. The matrix U will be left and it must satisfy (1.6.67) and (1.6.68). Evidently then, this is the only change in moving from the square to the nonsquare case.

Nehari's Theorem

Nehari's Theorem is an important special case of the optimal approximation result, corresponding to $r = 0$. The result is as follows:

Theorem 1.6.5. *Suppose $G \in RH_\infty$. Then*

$$\|G\|_H = \min_{F \in RH_\infty^-} \|G - F\|_\infty. \tag{1.6.78}$$

If G has degree n, then F can be chosen with degree $n - l$, where l is the multiplicity of the largest singular value of G. Further

$$\sigma_i\left[F(-s)\right] \le \sigma_{i+l}\left[G(s)\right], \qquad i = 1, \ldots, n-l, \tag{1.6.79}$$

with equality for square G.

The only part of this theorem that has not been established is the inequality (1.6.79). It is slightly easier to prove the result for square $G(s)$, and we shall do this. The formulae (1.6.55) through (1.6.58) become formulae for a realization of $F(s)$ once $r = 0$. One can verify that

$$E_1^{-1}Q_1\hat{A}^T + \hat{A}\left(E_1^{-1}Q_1\right) + \hat{B}\hat{B}^T = 0 \tag{1.6.80}$$

and

$$P_1E_1\hat{A} + \hat{A}^T\left(P_1E_1\right) + \hat{C}^T\hat{C} = 0. \tag{1.6.81}$$

Hence the Hankel eigenvalues of $F(-s)$ are the eigenvalues of $P_1E_1E_1^{-1}Q_1 = P_1Q_1$, i.e., $\sigma_{l+1}(G), \sigma_{l+2}(G), \ldots, \sigma_n(G)$.

The class of all optimal Hankel norm approximants of a nonsquare transfer function matrix

Let $G(s)$ be $p \times m$, stable, of degree n and strictly proper. The last assumption is for convenience; any D term is not part of the Hankel norm approximation problem for $G(s)$. If D is present, then it is simply added to the approximant $G_r(s) + F(s)$ of $G(s)$.

The previous subsection has shown that an optimal approximant $Q(s)$ can be constructed using certain formulas which ensure that the stable part has degree r; the formulas involve a matrix U subject to certain side constraints. The formulas are provided in equations (1.6.55) through (1.6.58), together with the requirement that U satisfy $U^T U \leq I$ and $B_2 = -C_2^T U$.

Define a transfer function matrix $R(s)$ by

$$\bar{A} = E_1^{-1}\left(\sigma_{r+1}^2 A_{11}^T + Q_1 A_{11} P_1\right), \tag{1.6.82}$$

$$\bar{B} = E_1^{-1}\begin{bmatrix} Q_1 B_1 & -\sigma_{r+1} C_1^T \end{bmatrix}, \tag{1.6.83}$$

$$\bar{C} = \begin{bmatrix} C_1 P_1 \\ -B_1^T \end{bmatrix}, \tag{1.6.84}$$

$$\bar{D} = \begin{bmatrix} 0 & \sigma_{r+1} I \\ I & 0 \end{bmatrix}. \tag{1.6.85}$$

It is a straightforward matter to verify that

$$Q(s) = \mathcal{F}_l\left[R, -\sigma_{r+1}^{-1} U\right]. \tag{1.6.86}$$

It turns out that the set of all $H(s) \in RH_\infty^-(r)$ which satisfy

$$\|G - H\|_\infty = \sigma_{r+1} \tag{1.6.87}$$

is given by a generalization of the formula for $Q(s)$:

$$H = \mathcal{F}_l\left[R, -\sigma_{r+1}^{-1} V\right], \tag{1.6.88}$$

where $V(s) \in RH_\infty^-$, $V_* V \leq I$ and $B_2 + C_2^T V(s) = 0$. This result can be found in Glover (1984) and Green and Limebeer (1995) and can be derived also by a variant on the argument for the suboptimal problem.

Some singular value inequalities

For future usc, wc want to record some inequalities relating Hankel singular values of certain quantities

We will show that for any stable $\hat{G}(s)$ of degree r, there holds

$$\sigma_i(G-\hat{G}) \geq \sigma_{i+r}(G), \qquad i = 1\ldots, n-r \tag{1.6.89}$$

(which we have already established for the case $i = 1$ in (1.6.18)) and

$$\sigma_{i+r}(G-\hat{G}) \leq \sigma_i(G), \qquad i = 1, \ldots, n. \tag{1.6.90}$$

For the first, let $\hat{H}$ of degree $i-1$ be such that

$$\sigma_1(G-\hat{G}-\hat{H}) = \sigma_i(G-\hat{G}) \tag{1.6.91}$$

[with $\hat{H}$ existing by the result on optimal Hankel norm approximation, see (1.6.64)]. Since $\hat{G}+\hat{H}$ has degree less than or equal to $i+r$, there holds

$$\sigma_1(G-\hat{G}-\hat{H}) \geq \sigma_{i+r}(G). \tag{1.6.92}$$

Equation (1.6.89) is immediate. Equation (1.6.90) follows by observing that

$$G = (G-\hat{G}) - (-\hat{G})$$

and replacing G and $\hat{G}$ in (1.6.89) by $G-\hat{G}$ and $-\hat{G}$ respectively.

Norm error bounds with stable approximants

In approximating an nth degree stable transfer function matrix $G(s)$ by an rth degree stable transfer function matrix $G_r(s)$ our main interest will be in the approximation error $\|G-G_r\|_\infty$. We have given almost no results on this error to this point; that is the focus of this subsection. Broadly, there are two approaches to the results, using different types of approximation.

The first approach depends on the following lemma.

Lemma 1.6.6. *Let $G(s)$ be a square stable transfer function matrix of degree n such that $\sigma_n(G)$ has multiplicity l. Let $\hat{G}(s)$ be the optimal Hankel norm approximation of degree $n-l$ corresponding to $F(s) \equiv 0$. Then*

1. $\sigma_n^{-1}(G)[G-\hat{G}]$ is an all-pass.

2.
$$\sigma_i(\hat{G}) = \sigma_i(G), \qquad i = 1, \ldots, n-l. \tag{1.6.93}$$

Proof. The first claim is immediate from the fact that $E = G - \hat{G} - F$ is a scaled all-pass, and $F \equiv 0$. For the second claim, notice that the construction of $\hat{A}$, $\hat{B}$, $\hat{C}$, $\hat{D}$ in (1.6.55) to (1.6.58) provides a realization for $\hat{G}$ in this case. Further,

$$(E_1^{-1}Q_1)\hat{A}^T + \hat{A}(E_1^{-1}Q_1) + \hat{B}\hat{B}^T = 0, \tag{1.6.94}$$

$$P_1E_1\hat{A} + \hat{A}^T(P_1E_1) + \hat{C}^T\hat{C}^T = 0, \tag{1.6.95}$$

as can readily be verified. Hence the Hankel singular values of $\hat{G}$ are $\lambda_i^{1/2}(P_1Q_1)$, or $\sigma_1(G), \ldots, \sigma_{n-l}(G)$. □

Let the distinct Hankel singular values of G be denoted by $\nu_1 > \nu_2 > \cdots > \nu_N$, where ν_i has multiplicity r_i. Denote the $\hat{G}$ obtained above as $\hat{G}_{N-1}(s)$ and let $\hat{G}_i(s)$ be the approximation of $\hat{G}_{i+1}(s)$ obtained by the above Lemma. Then we see that

$$\hat{G}_{i+1}(s) - \hat{G}_i(s) = \nu_{i+1} E_{i+1}(s), \tag{1.6.96}$$

for some all-pass $E_{i+1}(s)$ and

$$G(s) - \hat{G}_k(s) = \nu_N E_N(s) + \cdots + \nu_{k+1}(s) E_{k+1}(s). \tag{1.6.97}$$

Hence this construction yields $\hat{G}_k$ of degree $r_1 + r_2 + \cdots + r_k$ such that

$$\left\| G(s) - \hat{G}_k(s) \right\|_\infty \leq \sum_{j=k+1}^{N} \nu_j. \tag{1.6.98}$$

This is arguably the most important H_∞ error formula associated with Hankel norm approximation. It extends trivially to nonsquare $G(s)$.

We see further by carrying the procedure through to $i = 0$ that

$$G(s) = D_0 + \sum_{i=1}^{N} \nu_i E_i(s), \tag{1.6.99}$$

for some constant D_0 and accordingly, there exists D_0 such that

$$\left\| G(s) - D_0 \right\|_\infty \leq \sum_{i=1}^{N} \nu_i. \tag{1.6.100}$$

Taking the strictly proper part of (1.6.97) also shows that if $G(s)$ and $\hat{G}_k(s)$ are strictly proper,

$$\left\| G(s) - \hat{G}_k(s) \right\|_\infty \leq \sum_{j=k+1}^{N} 2\nu_j \tag{1.6.101}$$

and

$$\left\| G(s) \right\|_\infty \leq \sum_{i=1}^{N} 2\nu_i. \tag{1.6.102}$$

For the second approach, we will rely on the following lemma.

Lemma 1.6.7. *Let*

$$E(s) = \left[\begin{array}{cc|c} A_1 & 0 & B_1 \\ 0 & A_2 & B_2 \\ \hline C_1 & C_2 & D \end{array}\right] = \left[\begin{array}{c|c} A & B \\ \hline C & D \end{array}\right] \tag{1.6.103}$$

be a square scaled all-pass, i.e., $E_*(s)E(s) = \sigma^2 I$. *Suppose that* $E(s)$ *has* $n_1 = \dim A_1$ *poles in* $\mathrm{Re}[s] < 0$ *and* $r_2 = \dim A_2$ *poles in* $\mathrm{Re}[s] > 0$, *the realization being minimal. Suppose* $n_1 > n_2$. *Let* $G(s) = C_1(sI - A_1)^{-1}B_1$ *and* $F(s) = D + C_2(sI - A_2)^{-1}B_2$. *Then*

$$\begin{aligned} \sigma_i(G) &= \sigma^2, & i &= 1, \ldots, n_1 - n_2, \\ &= \sigma_{i-(n_1-n_2)}\left[F(-s)\right], & i &= n_1 - n_2 + 1, \ldots, n_1. \end{aligned} \tag{1.6.104}$$

Proof. Let P, Q be the solutions of

$$PA^T + AP + BB^T = 0, \qquad QA + A^TQ + C^TC = 0. \tag{1.6.105}$$

Then the all-pass property ensures that $PC^T + BD^T = 0$ and thence that $Q = \sigma^2 P^{-1}$. Let

$$P = \begin{bmatrix} P_{11} & P_{12} \\ P_{12}^T & P_{22} \end{bmatrix}, \qquad Q = \begin{bmatrix} Q_{11} & Q_{21}^T \\ Q_{21} & Q_{22} \end{bmatrix}.$$

Then

$$\begin{aligned} \det(\lambda I - P_{11}Q_{11}) &= \det\left(\lambda I - \left(\sigma^2 I - P_{12}Q_{21}\right)\right) \\ &= \det\left[\left(\lambda - \sigma^2\right) I + P_{12}Q_{21}\right] \\ &= \left(\lambda - \sigma^2\right)^{n_1-n_2} \det\left[\left(\lambda - \sigma^2\right) I + Q_{21}P_{12}\right] \\ &= \left(\lambda - \sigma^2\right)^{n_1-n_2} \det\left[\lambda I - Q_{22}P_{22}\right]. \end{aligned} \tag{1.6.106}$$

Since $\sigma_i(G) = \lambda_i(P_{11}Q_{11})$ and $\sigma_i[F(-s)] = \lambda_i(Q_{22}P_{22})$ the result is immediate.

Suppose now that $E(s) = G(s) - G_r(s) - F(s)$, where $G_r(s) + F(s)$ has realization $\{\hat{A}, \hat{B}, \hat{C}, \hat{D}\}$ as defined in (1.6.55) to (1.6.58) and $\sigma = \sigma_{r+1}(G)$, which is assumed to have multiplicity l. The realization $\{A_e, B_e, C_e, D_e\}$ of E which we need is indeed controllable since we exhibited a nonsingular P_e for which $P_eA_e^T + A_eP_e + B_eB_e^T = 0$; in a similar manner, we can establish it is observable. (The argument is not completely trivial.) So it is minimal. The integer n_1 is $n+r$ and n_2 is $n-r-l$; thus $n_1 - n_2 = 2r+l$. By the lemma, we have

$$\sigma_i(G - G_r) = \sigma_{r+1}(G), \qquad i = 1, 2, \ldots, 2r + l, \tag{1.6.107}$$

$$\sigma_i(G - G_r) = \sigma_{i-(2r+l)}\left[F(-s)\right], \qquad i > 2r + l. \tag{1.6.108}$$

Using (1.6.90), we have

$$\sigma_i\left(G - G_r\right) \leq \sigma_{i-r}(G), \qquad i > 2r + l. \tag{1.6.109}$$

These inequalities are of interest in their own right. Also, by combining these inequalities with (1.6.100), we conclude that there exists D_0 for which

$$\begin{aligned} \|G - G_r - D_0\|_\infty &\le \sigma_{r+1}(G) + \sum_{i=2r+l+1}^{n+r} \sigma_i\,(G - G_r) \\ &\le \sigma_{r+1}(G) + \sum_{i=r+l+1}^{n} \sigma_i(G). \end{aligned} \tag{1.6.110}$$

This bound is not as attractive as that at (1.6.98), since the sum in (1.6.110) count multiple occurrences of the one singular value, in contrast to (1.6.98). Also, if one insists that G_r is to be strictly proper and no D_0 is permitted, application of (1.6.102) to $G - G_r$ yields

$$\|G - G_r\| \le 2\left[\sigma_{r+1}(G) + \sum_{i=r+l+1}^{n} \sigma_i(G)\right]. \tag{1.6.111}$$

Of course, the inequalities (1.6.109)–(1.6.111) are also valid for nonsquare G. □

In several of the bounds above, there appears a constant term D_0. In general, it is possible to optimize D_0 to minimize the norm of the relevant function in which it appears. This is a convex optimization problem. One may well be able to do better than the bounds as a result.

Connection with approximation by balanced truncation

Let $G(s)$ be square, stable, of degree n and with minimal realization $\{A, B, C\}$. Suppose that the realization is balanced. Let $\{\hat{A}, \hat{B}, \hat{C}, \hat{D}\}$ be the realization of G_{n-l} constructed in equations (1.6.55) through (1.6.58). Here l is the multiplicity of the smallest Hankel singular value of G. Note that the matrices E_1, P_1 and Q_1 are all diagonal, and the latter two are equal. Let $G_b(s)$ denote the degree $n - l$ transfer function matrix of the truncation of the balanced realization of $G(s)$, where only the last singular value is thrown away. Then it is not difficult to check that $\sigma_n^{-1}(G_b - G_{n-l})$ is an all-pass. It follows easily that

$$\|G - G_b\|_\infty \le \|G - G_{n-l}\|_\infty + \|G_{n-l} - G_b\|_\infty = 2\sigma_n.$$

This is an alternative derivation of the error formula for one step of a balanced truncation, and generalizes obviously to a truncation where more than one Hankel singular value is thrown away.

Main points of the section

1. The Hankel operator is an operator defined from the impulse response of the system which maps past inputs to future outputs. The Hankel norm is the induced L_2 norm of the operator.

2. The singular values of a Hankel operator of a finite dimensional system are the square roots of the eigenvalues of the product of the controllability and observability grammians, and the maximum singular value is the Hankel norm.

3. There is a stable G_r of McMillan degree r or less such that $\|G - G_r\|_H < \gamma$ if $\gamma > \sigma_{r+1}(G)$. Reduced order models G_r satisfying the norm condition constitute the class of suboptimal approximations of G. The class of all suboptimal approximants can be expressed using a Lower Fractional Transformation formula.

4. Optimal Hankel norm approximations achieve the lower error bound, which is $\sigma_{r+1}(G)$. The class of all optimal approximants can be expressed using a Lower Fractional Transformation formula: G_r = stable part of $F_l(R, -\sigma_{r+1}^{-1}V)$, where R is defined in a standard way using the realization of G and σ_{r+1}, and V satisfies $V \in RH_\infty^-$, $V_* V \leq I$ and a condition of the form $B_2 + C_2 V(s) = 0$, where B_2 and C_2 are constant matrices derivable from the realization of $G(s)$.

5. The Nehari theorem constitutes a special case of optimal Hankel norm approximation: $\|G\|_H = \min_{F \in RH_\infty^-} \|G - F\|_\infty$.

6. An upper bound on the H_∞ norm of the error $G - G_r$ is known, when G_r is an optimal Hankel norm approximation. The bound is the sum of the distinct truncated singular values (*i.e.*, $\sigma_{r+1} + \cdots + \sigma_n$, excluding repeating values), which is half the bound for the balanced truncation case.

1.7 Markov and Covariance Parameter Matching

Padé approximation and Markov parameter matching

It is reasonable that reduced order models keep the same characteristics partially as the original systems, and since Markov parameters characterize system dynamics, it is reasonable to consider the idea of matching such parameters as a basis for model reduction. Equally, one could consider working with a Taylor-series expansion of $G(s)$. Thus, expanding a stable transfer function $G(s)$ in powers of s yields

$$G(s) = \sum_{i=0}^{\infty} g_i s^i. \tag{1.7.1}$$

The coefficients have the following relation with the impulse response $g(t)$.

$$g_i = (-1)^i \frac{1}{i!} m_i, \tag{1.7.2}$$

where $m_i = \int_0^\infty t^i g(t)dt$ is called the ith order time moment. One can obtain a reduced order model $G_r(s)$ which has the same m_i up to a certain order, Zakian (1973).

If the obtained reduced order model is stable and rational, the model is called a *Padé* approximation. The number of matched coefficients is dependent upon the order of the model. In the case of single input-single output systems, we have the relation:

$$\int_0^{\infty} t^i \left(g(t) - g_r(t)\right) dt = 0, \qquad i = 0, 1, \ldots, n_r + m_r, \tag{1.7.3}$$

where n_r and m_r are the order of the denominator and numerator of $G_r(s)$, respectively. It can be seen from this relation that the obtained model approximates the original impulse response in the low frequency range. It is also possible to expand $G(s)$ in negative powers of s instead of (1.7.1):

$$G(s) = \sum_{i=0}^{\infty} h_i s^{-i}. \tag{1.7.4}$$

The coefficients are called Markov parameters, and they also characterize the system response. The matching of these parameters leads to better approximation in the high frequency range. (This approximation procedure is sometimes termed partial realization approximation.)

A method mixing these two techniques has been proposed to obtain goodness of matching in both low and high frequency ranges, Shamash (1975b). The calculation is virtually the same as that applying when one constructs a minimal realization from a finite set of Markov parameters or time moments, and is easy; however, a fundamental problem is that there is no guarantee of stability of the reduced order model. That is, the reduced order model can be either stable or unstable although the original system is asymptotically stable.

Similar methods exist using continued fraction expansions. The truncation of a continued fraction expansion of a high order model leads to the reduced order model. There normally exist two continued fraction expansions for which the reduced model is exactly the same as that resulting from Markov parameter or time moment matching. There are however some situations where an expansion is not possible; thus, this method is restrictive, Wright (1973), Calfe and Healey (1974).

Antoulas and Bishop (1987) extended this use of continued fraction expansion to multiple-input, multiple-output systems, with truncation of the continued fraction yielding a reduced order model. The ideas depend on the concept of partial realization; again, the stability problem cannot be straightforwardly addressed.

Despite there being no clear connection between the stability of the original and the reduced order model, there is one nice property flowing from using partial matching of Markov or time-moment parameters, which is summarized in the following lemma, Obinata (1978).

Lemma 1.7.1. *Consider a composite system S built up from three basic connection types of parallel, tandem and feedback from a set of subsystems S_i. Then the jth moment of S is determined only by the moments $m_{i,k}$ $(k = 0, 1, \ldots, j)$, where $m_{i,k}$ is the kth moment of the subsystem S_i.*

It follows from this Lemma that if a low order approximation model of a plant matches the time moments $m_0, \ldots, m_j$, a controller defined on the basis of the low order model will, when used in closed loop with the original plant, give the same closed-loop time moments as it would in closed-loop with the low order model. This property is also true for Markov parameter matching.

The relation between open-loop and closed-loop approximation is the key theme of later chapters.

Covariance parameter matching and the relation to equation error techniques

A power spectrum is usually a property of a signal; however, the concept is also applicable to systems. A power spectrum $\Phi(s)$ can be defined using a transfer function $G(s)$ as

$$\Phi(s) = G(s)G_*(s). \tag{1.7.5}$$

This can be decomposed with stable $Z(s)$ and unstable $Z_*(s)$ as follows:

$$\Phi(s) = Z(s) + Z_*(s). \tag{1.7.6}$$

Expanding $Z(s)$ in negative powers of s yields

$$Z(s) = \frac{1}{2}z_0 + \sum_{i=1}^{\infty} z_i s^{-i}. \tag{1.7.7}$$

The coefficients z_i are termed covariance parameters, and characterize the power spectrum. Matching these parameters gives other methods for model reduction. One method, termed the q-Markov COVER method, matches not only covariance parameters but also Markov parameters, Yousuff *et al.* (1985). If the original system is asymptotically stable, we can always find a coordinate basis for the state space such that the controllability grammian is equal to the unit matrix and the state equation is given by

$$\begin{bmatrix} \dot{x}_1 \\ \dot{x}_2 \\ \vdots \\ \dot{x}_p \end{bmatrix} = \begin{bmatrix} A_{1,1} & A_{1,2} & 0 & \cdots & 0 \\ A_{2,1} & A_{2,2} & A_{2,3} & \ddots & \vdots \\ \vdots & \vdots & \vdots & \ddots & 0 \\ A_{p-1,1} & A_{p-1,2} & A_{p-1,3} & \cdots & A_{p-1,p} \\ A_{p,1} & A_{p,2} & A_{p,3} & \cdots & A_{pp} \end{bmatrix} \begin{bmatrix} x_1 \\ x_2 \\ \vdots \\ x_p \end{bmatrix} + \begin{bmatrix} B_1 \\ B_2 \\ \vdots \\ B_p \end{bmatrix} w, \tag{1.7.8}$$

$$y = \begin{bmatrix} C_1 & 0 & 0 & \ldots & 0 \end{bmatrix} x. \tag{1.7.9}$$

The dimension of x_i is n_i and

$$\sum_{i=1}^{p} n_i = n, \qquad \operatorname{rank}(A_{i,i+1}) = n_{i+1} \le n_i, \qquad i = 1, 2, \ldots, p-1. \tag{1.7.10}$$

The observability indices are defined as $r_i = \text{rank}(R_{0i}^T)$ where

$$R_{0i}^T = \left[C^T, A^T C^T, \ldots, (A^T)^{i-1} C^T\right]$$

and the indices n_i in (1.7.8) are related to the observability indices r_i.

Truncating (1.7.8) to eliminate x_j with $j > q$ (as per Section 2) yields a reduced order model of dimension r_q. The model has the property that the first q covariance parameters match those of the original system. In addition, the following three properties hold.

1. Provided the reduced order model is controllable, all eigenvalues of the reduced "A" matrix lie in $\text{Re}[s] < 0$; otherwise, some may be on the imaginary axis.

2. The controllability grammian of the reduced order system is again the unit matrix.

3. The first q Markov parameter of the reduced order model and the original system are the same.

There is a connection between this method and the L_2 error norm minimization approach to low order approximation of Section 1.5. Mullis and Roberts (1976), Inouye (1983) and Obinata (1989) display the connection between minimizing the error norm in approximating a discrete-time transfer function and the above covariance and Markov parameter matching. The covariance parameters are associated with system steady state behaviour, and so matching these at the same time as one matches Markov parameters provides good approximation in both the low and high frequency ranges, which may be desirable in many model reduction problems.

In the above methods, the number of covariance and Markov parameters is required to be the same; de Villemagne and Skelton (1987) have proposed a method allowing matching of (independently) related numbers of these parameters or generalized parameters, using projection. Also, rather than viewing the problem as one of model reductions, it can be viewed as an interpolation problem, Horiguchi, Nishimura and Nagata (1990).

Main points of the section

1. Methods for model reduction based on matching of time moments or Markov parameters have been proposed. The methods do not guarantee, even if the original system is stable, that the reduced order system will be stable.

2. Matching of time moments or Markov parameters in open loop implies matching in closed loop.

3. A method for matching both Markov parameters and covariance parameters has been proposed. It has a close relation to equation error methods in identification.

1.8 Mixed Methods

Model reduction using Linear Matrix Inequalities (LMIs)

Using an expression for the additive error $E(s)$, we can consider the suboptimal problems of finding a reduced model such that

$$\|E(s)\|_\infty < \gamma \quad \text{or} \quad \|E(s)\|_2 < \gamma, \tag{1.8.1}$$

where

$$E(s) = G(s) - G_r(s) = \left[\begin{array}{c|c} \tilde{A} & \tilde{B} \\ \hline \tilde{C} & \tilde{D} \end{array}\right] = \left[\begin{array}{cc|c} A & 0 & B \\ 0 & A_r & -B_r \\ \hline C & C_r & D - D_r \end{array}\right]. \tag{1.8.2}$$

It is assumed that $G(s)$ and $G_r(s)$ are both stable transfer functions. Then the first condition of (1.8.1) can be rewritten using the Bounded Real Lemma as follows, see Green and Limebeer (1995). There exists $\tilde{P} \geq 0$ such that

$$\begin{bmatrix} \tilde{A}^T \tilde{P} + \tilde{P} \tilde{A} & \tilde{P} \tilde{B} & \tilde{C}^T \\ \tilde{B}^T \tilde{P} & -\gamma I & \tilde{D}^T \\ \tilde{C} & \tilde{D} & -\gamma I \end{bmatrix} < 0. \tag{1.8.3}$$

The second condition can be rewritten as follows. There holds $D = D_r$ and there exists $\breve{P} \geq 0$ such that

$$\begin{bmatrix} \tilde{A} \breve{P} + \breve{P} \tilde{A}^T & \tilde{B} \\ \tilde{B}^T & -I \end{bmatrix} < 0, \qquad \operatorname{tr}(\tilde{C} \breve{P} \tilde{C}^T) < \gamma^2. \tag{1.8.4}$$

By partitioning $\tilde{P}$ or $\breve{P}$ as

$$\tilde{P} = \begin{bmatrix} P_{11} & P_{12} \\ P_{21} & P_{22} \end{bmatrix} \quad \text{or} \quad \breve{P} = \begin{bmatrix} \breve{P}_{11} & \breve{P}_{12} \\ \breve{P}_{21} & \breve{P}_{22} \end{bmatrix}$$

we can rewrite (1.8.3) and (1.8.4) as

$$\begin{bmatrix} A^T \tilde{P}_{11} + \tilde{P}_{11} A & A^T \tilde{P}_{12} + \tilde{P}_{12} A_r & \tilde{P}_{11} B - \tilde{P}_{12} B_r & C^T \\ A_r^T \tilde{P}_{12}^T + \tilde{P}_{12}^T A & A_r^T \tilde{P}_{22} + \tilde{P}_{22} A_r & \tilde{P}_{12}^T B - \tilde{P}_{22} B_r & C_r^T \\ B^T \tilde{P}_{11} - B_r^T \tilde{P}_{12}^T & B^T \tilde{P}_{12} - B_r^T \tilde{P}_{22} & -\gamma I & D^T - D_r^T \\ C & C_r & D - D_r & -\gamma I \end{bmatrix} < 0 \tag{1.8.5}$$

and

$$\begin{gathered} \begin{bmatrix} A \breve{P}_{11} + \breve{P}_{11} A^T & A \breve{P}_{12} + \breve{P}_{12} A_r^T & B \\ A_r \breve{P}_{12}^T + \breve{P}_{12}^T A^T & A_r \breve{P}_{22} + \breve{P}_{22} A_r^T & B_r \\ B^T & B_r^T & -I \end{bmatrix} < 0, \\ \operatorname{tr}(C \breve{P}_{11} C^T + 2 C \breve{P}_{12} C_r^T + C_r \breve{P}_{22} C_r^T) < \gamma^2. \end{gathered} \tag{1.8.6}$$

The approach adopted is to find the smallest possible γ with respect to A_r, B_r, C_r, D_r based on solutions of the suboptimal problems. The optimization problem is not linear in its unknown parameters, since bilinear terms appear in the inequalities. If A_r, B_r or A_r, C_r are fixed at certain values however, then the matrix inequalities (1.8.5) and (1.8.6) are linear in C_r, D_r or B_r, D_r and $\tilde{P}$ and $\breve{P}$. The idea from this point is to mix two methods, to exploit this last observation. First, obtain A_r, B_r or A_r, C_r using mode truncation or balanced truncation or another reduction method. Second, optimize C_r, D_r or B_r, D_r based on the linear matrix inequality (1.8.5) or (1.8.6). We can enjoy the nice properties of convex optimization techniques in this step; thus, the convergence to the smallest possible γ (consistent with the A_r, B_r or A_r, C_r pair) can be achieved, Helmersson (1994).

The most simple method in this category is to combine the state aggregation method and 2-norm minimization. The feedthrough term D_r should be taken as D as in the case of normal 2-norm minimization. A_r, B_r are given from (1.2.12) as follows:

$$A_r = LAL^T(LL^T)^{-1}, \qquad B_r = LB. \tag{1.8.7}$$

Since (1.2.11) holds in aggregated models, minimizing $\|E(s)\|_2$ leads to the following explicit expression of C_r:

$$C_r = C\breve{P}_{11}L^T(L\breve{P}_{11}L^T)^{-1} \tag{1.8.8}$$

and we can actually take $\breve{P}_{11}$ as the controllability grammian of the pair (A, B), see Inooka and Obinata (1977).

Mixed method with singular perturbation and balanced realization

In Section 1.4, reduced order models are obtained by truncating the state space expressions of balanced realizations. However, as we know, to further reduce the order, we can also apply singular perturbation to the balanced realization.

Assume that A, B, C, D is the minimal and internally balanced realization of an asymptotically stable system, with controllability and observability grammian Σ. Then the singular perturbation reduced-order model defined as in (1.3.5) and (1.3.6) is also asymptotically stable and controllable and observable if and only if the submatrices Σ_1 and Σ_2 of Σ (as defined in (1.4.11)) have no common diagonal element. The proof is immediate from the fact that the 'reciprocal system' $A^{-1}, A^{-1}B, CA^{-1}, D$ is also stable and internally balanced with the same grammian if the original system is stable and internally balanced, Liu and Anderson (1989). Moreover, the same error bound as in the case of balanced truncation is available; that is, (1.4.12) holds with this reduced order model.

We have seen in Section 1.2 and Section 1.3 that if we use truncation on a balanced system to obtain a reduced order approximation, we have a small reduction error at very high frequencies, but not necessarily such a good one at low frequencies. If we use the singular perturbation technique, we have the reverse conclusion. Hence we come to the idea of mixing these two reduction techniques on a stable balanced system for averaging the behaviour of the two methods. If we make a reduction through several sequential steps, truncation or singular perturbation technique can be used for each step. (Liu and Anderson, 1989) proved that the final reduced order model is minimal, balanced and stable; further, the error bound will remain the same as if the reduction had been done by either method in one step. The frequency distribution of the actual error of course cannot be expected to be the same.

Other mixed methods

Many types of mixed methods for model reduction can be considered for combined use in a particular reduction problem. The aggregation method or mode truncation is especially easy to use with other methods. The most simple application of this method is to use mode truncation just for obtaining A_r or the denominator of the reduced order transfer function. Then B_r, C_r, D_r or the coefficients in the numerator are calculated using another method. The selection of A_r or the denominator by mode truncation will give a large simplification to the step for obtaining B_r, C_r, D_r or the coefficients of the numerator. This technique also solves the stability problem arising with *Padé* approximation or Markov parameter matching; that is, we can always obtain stable reduced order models from a stable original system, even in the case when *Padé* approximation or the Markov parameter matching technique gives an unstable reduced order model, (Shamash, 1975a).

Main points of the section

1. Suboptimal solutions for finding a stable reduced order model $\{A_r, B_r, C_r, D_r\}$ to approximate a stable model in the H_2 or H_∞ norm sense are easily obtained if the pair A_r, B_r or the pair A_r, C_r is first obtained. Linear matrix inequalities, solvable by convex optimization technique, are involved.

2. Several mixed reduction methods can be proposed. One involves mixing truncations and singular perturbation of a balanced realization. At each sequential step, either truncation or singular perturbation occurs. There is a common H_∞ norm error bound.

1.9 Model Reduction in Discrete Time Systems

Bilinear transformation

Any stable discrete time transfer function $G_d(z)$ can be used to define a stable continuous time transfer function by the following bilinear transformation:

$$G(s) = G_d\left(\frac{\alpha + s}{\alpha - s}\right), \tag{1.9.1}$$

where α is a scalar parameter. Using the transformation, we can obtain reduced order discrete time models from reduction techniques for continuous time models. First, apply the transformation (1.9.1) to the original discrete time transfer function $G_d(z)$. Second, reduce $G(s)$ by any continuous time method to obtain $G_r(s)$. Finally, apply the following reverse transformation to $G_r(s)$ to obtain the reduced order discrete time model:

$$G_{dr}(z) = G_r\left(\alpha\frac{z-1}{z+1}\right). \tag{1.9.2}$$

The Hankel singular values of $G_d(z)$ or $G_{dr}(z)$ are the same as those of $G(s)$ or $G_r(s)$, as it turns out. Therefore, the H_∞ error bound remains the same when Hankel optimal approximation is used for the continuous time reduction. The selection of α affects the final result; there is however no simple rule for choosing α.

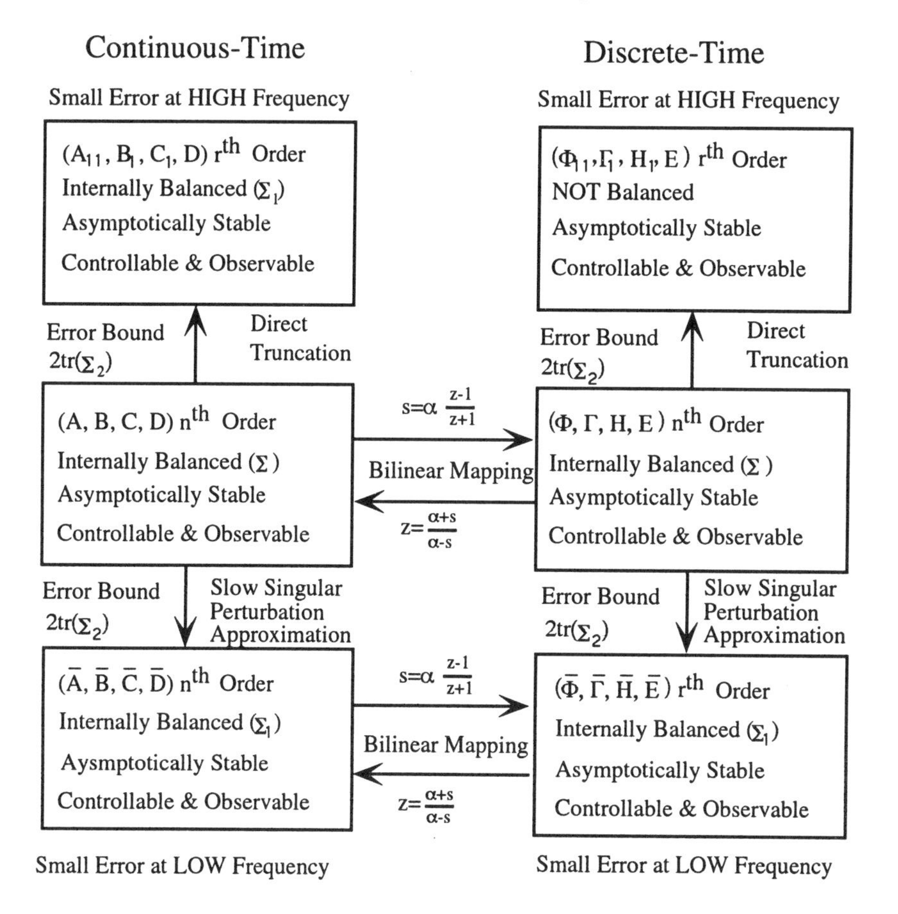

Figure 1.9.1. Relationship of truncation and singular perturbation of a balanced realization for continuous and discrete time cases

It is also possible to perform directly truncation of a balanced discrete-time realization, Al-Saggaf and Franklin (1986). An error formula is available, but the truncated realization is no longer balanced. Likewise, singular perturbation of a balanced discrete-time realization can be performed; the same error formula applies, and this time, the reduced order model is balanced. (Thus general mixing of truncation and singular perturbation steps is not permissible, while singular perturbation followed by truncation is permissible.) Figure 1.9.1 drawn from Liu and Anderson (1989) summarizes a number of the ideas. Note that in the figure, $G(s) = D + C(sI - A)^{-1}B$ and $G_d(z) = E + H(zI - \Phi)^{-1}\Gamma$, with the state-variable realization internally balanced. The subscript 1-1 applies to truncation and the overbar to a singular perturbation approximation. The figure shows that bilinear transformation followed by balanced realization singular perturbation reduction and balanced realization singular perturbation reduction followed by bilinear transformation lead to the same thing. (An equivalent statement does not hold true for truncation and bilinear transformation.) Liu and Anderson (1989) discusses these issues in some detail.

Approximation by finite impulse response transfer functions

Finite impulse response (FIR) transfer functions are widely used because they have good numerical properties and are easy to implement. Approximating infinite impulse response (IIR) transfer functions by FIR transfer functions is a kind of model reduction. (It may be better to use the term 'simplification' than 'reduction' for this case since the state-variable dimension may well increase.)

We can apply ideas similar to some introduced for the continuous time case to this problem. For example, if we apply a Markov parameter matching technique to a discrete time IIR transfer function, then the simple truncation retaining lower order coefficients is obtained as the approximation. Least squares minimization techniques are also applicable to FIR approximation, and are widely used in signal processing or identification problems. Here, let us simply note an extension of the 2-norm minimization technique to allow for frequency weighting. (The idea has relevance in later chapters.) The problem is to find a FIR $\hat{G}(z)$ with r coefficients such that

$$\min_{\hat{g}} \left\| W(z)\left[G(z) - \hat{G}(z)\right] \right\|_2, \tag{1.9.3}$$

where $W(z)$ is a frequency weighting (or can be considered as a prefilter for the error system). Of course $G(z)$ is the IIR transfer function to be approximated, and $\hat{g}$ is the vector consisting of the coefficients of $\hat{G}(z)$, that is

$$\begin{aligned} \hat{G}(z) &= g_0 + g_1 z^{-1} + g_2 z^{-2} + \cdots + g_{r-1} z^{-r+1}, \\ \hat{g} &= \begin{bmatrix} g_0 & g_1 & g_2 & \cdots & g_{r-1} \end{bmatrix}^T. \end{aligned} \tag{1.9.4}$$

The approximation problem can be solved if both $W(z)$ and $G(z)$ are asymptotically stable, see Obinata *et al.* (1988); further, the solution is given from the following linear equation:

$$\begin{bmatrix} \phi_0 & \phi_1 & \dots & \phi_{r-1} \\ \phi_1 & \phi_0 & \dots & \phi_{r-2} \\ \vdots & \vdots & \ddots & \vdots \\ \phi_{r-1} & \phi_{r-2} & \dots & \phi_0 \end{bmatrix} \begin{bmatrix} g_0 \\ g_1 \\ \vdots \\ g_{r-1} \end{bmatrix} = \begin{bmatrix} \rho_0 \\ \rho_1 \\ \vdots \\ \rho_{r-1} \end{bmatrix}, \tag{1.9.5}$$

where

$$\rho_i = \sum_{j=0}^{\infty} m_{j+i} w_j, \qquad \phi_i = \sum_{j=0}^{\infty} w_j w_{j+i}, \qquad i = 0, 1, \dots, r-1 \tag{1.9.6}$$

and where $m_i, i = 0, \dots, \infty$ is the impulse response sequence of $W(z)G(z)$ and $w_i, i = 0, \dots, \infty$ is the impulse response sequence of $W(z)$. The ρ_i and ϕ_i can be calculated through discrete time Lyapunov equations when $W(z)$ and $G(z)$ are rational. All calculations required are linear; thus, it is very easy to get the solutions. Moreover, since the matrix in (1.9.5) is a Toeplitz matrix, it is easy to study the variation of the resulting error norm with the order of the FIR transfer function.

Nehari Shuffle

In the case of approximation with a scalar FIR transfer function, we can obtain a very sophisticated technique using the H_∞ error norm. The technique has been termed the Nehari Shuffle by the originators, Kootsookos, Bitmead and Green (1992). The error criterion used is

$$\left\| E(\omega) \right\|_\infty = \max_{\omega \in (-\pi, \pi]} \left| G(e^{j\omega}) - \hat{G}(e^{j\omega}) \right|, \tag{1.9.7}$$

where $G(z)$ is an IIR transfer function to be approximated by a FIR transfer function $\hat{G}(z)$. Assume that the McMillan degree of the unknown $G(z)$ is n and the number of coefficients of $\hat{G}(z)$ is r.

Now observe the following decomposition $G(z)$ is always possible.

$$G(z) = \sum_{i=0}^{r-1} g_i z^{-i} + z^{-(r-1)} \sum_{j=1}^{\infty} g_{j+r-1} z^{-j} = G^h + z^{-(r-1)} G^l. \tag{1.9.8}$$

Then we define the following operators:

$\aleph G$: Nehari extension of G (see later)

$\hbar G$: Extraction of G^h from G.

$\Im G$: Extraction of the tail G^l from G. When G has realization $\{A, B, C, D\}$, the realization of $\Im G$ is given by $\{A, B, CA^{r-1}, 0\}$.

$\Re G$: Reflection operator $\Re : G(z) \to G(z^{-1})$. The realization of $\Re G$ is $\{A^{-1}, A^{-1}B, -CA^{-1}, D - CA^{-1}B\}$.

∇G: Shift operation $\nabla : G \to z^{-(r-1)}G$. $\nabla^{-1}G$ denotes $z^{r-1}G$.

In Section 1.6 (Hankel Norm approximation), the Nehari theorem was introduced. It can be applied to the discrete time case as follows:

Theorem 1.9.1 (Discrete-time Nehari Theorem). *Let $G(z)$ be a real rational function which is analytic in $\{|z| > \rho\}$ for some $\rho < 1$. Then there exists a real rational $F(z)$ (the Nehari extension of $G(z)$) constrained to be analytic in $\{|z| < \eta\}$ for some $\eta > 1$ which minimizes* $\max_{\omega\in(-\pi,\pi]}|G(e^{j\omega}) - F(e^{j\omega})|$.

We can obtain the Nehari extension $F(z)$ using a bilinear transformation as described earlier in this section, and optimal Hankel norm approximation as described in Section 1.6. $F(z)$ is the closest anticausal transfer function to the causal $G(z)$. The McMillan degree of $F(z)$ is $n-1$. This implies that the operator $\aleph$ reduces the degree of the operand by one. Further we know from the discussion of Section 1.6 that

$$\sigma_i(\Re F) = \sigma_{i+1}(G), \quad \text{for} \quad i = 1, \ldots, n-1 \tag{1.9.9}$$

and that

$$\max_{\omega\in(-\pi,\pi]} \left|G(e^{j\omega}) - F(e^{j\omega})\right| = \bar{\sigma}(G). \tag{1.9.10}$$

This theorem is the basis of the Nehari Shuffle.

We shall first derive a lower bound on the approximation error between G and any FIR $\hat{G}$. The following equality can be easily obtained from the definitions:

$$\begin{aligned} \left|G(e^{j\omega}) - \hat{G}(e^{j\omega})\right| &= \left|\hbar G(e^{j\omega}) + \nabla\Im G(e^{j\omega}) - \hat{G}(e^{j\omega})\right| \\ &= \left|\nabla\Im G(e^{j\omega}) - \tilde{G}(e^{j\omega})\right|, \end{aligned}$$

where $\tilde{G} = \hat{G} - \hbar G$. Observe that $\Im G$ is strictly causal and $\nabla^{-1}\tilde{G}$ is anticausal. Use the fact that for every transfer function X, $|\nabla X| = |X|$ on $z = e^{j\omega}$, and apply the Nehari Theorem to $\Im$. Since $\nabla^{-1}\tilde{G}(e^{j\omega})$ is not necessarily the Nehari approximation, we conclude that

$$\min_{\tilde{G}} \max_{\omega\in(-\pi,\pi]} \left|\Im G(e^{j\omega}) - \nabla^{-1}\tilde{G}(e^{j\omega})\right| \geq \bar{\sigma}(\Im G) \tag{1.9.11}$$

It follows that

$$\min_{\hat{G}} \max_{\omega\in(-\pi,\pi]} \left|G(e^{j\omega}) - \hat{G}(e^{j\omega}\right| \geq \bar{\sigma}(\Im G). \tag{1.9.12}$$

Notice that there is no suggestion that this lower bound is attainable with some FIR $\hat{G}$. The lower bound is of course easily computable, depending as it does on $\Im G$ for which a state-variable realization is available.

Before formally stating the approximation algorithm, we shall first explain it, with the aid of Figure 1.9.2. The transfer function of the IIR system we wish to approximate is $G(z)$. We will use transfer functions now to label the associated impulse responses. Since the algorithm contains a series of approximation steps, we will set $G_1(z) = G(z)$. This is the bold line in the figure. The first and crudest FIR approximation to $G_1(z)$ is the r-coefficient truncation $\hat{G}_1(z)$ (which overlaps $G_1(z)$ through time $r-1$ and is zero thereafter). We need to improve the approximation by accounting somehow for the tail of $G_1(z)$. Accordingly, we perform a Nehari extension of $\Im G_1$, and shift it right by r time units. We denote the Nehari extension as $G_2(z^{-1})$, and its right shifting is then $z^{-(r-1)}G_2(z^{-1})$. Evidently, $G_2 = \Re\aleph\Im G_1$.

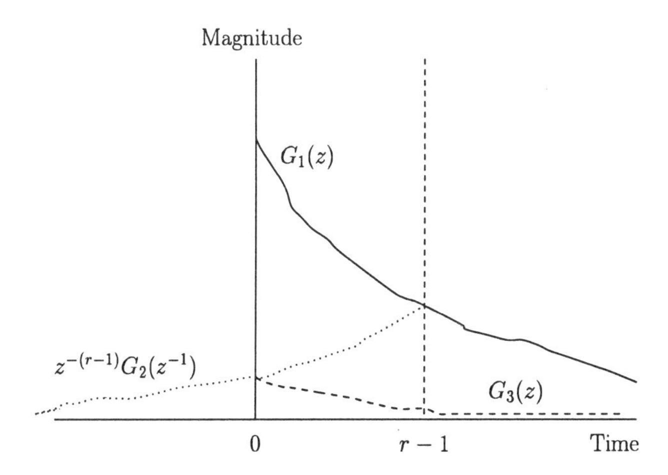

Figure 1.9.2. Illustration of the Nehari shuffle algorithm

The part of $z^{-(r-1)}G_2(z^{-1})$ between 0 and $r-1$ can contribute to improvement of $\hat{G}_1(z)$ as an FIR approximation of G_1. However, the anticausal part cannot. Nevertheless, we can perform a Nehari extension of the anticausal part of $z^{-(r-1)}G_2(z^{-1})$ to use this as a replacement; this second extension, $G_3(z)$, is the dashed line in the figure. That part of $G_3(z)$ between 0 and $r-1$ can be added to the approximant of $G_1(z)$, and then the process is repeated on the tail of $G_3(z)$.

The algorithm proceeds as follows.

1. Set $i := 1$, $\hat{G}_i := \hbar G$, and $G_i := G$.

2. Repeat

$$G_{i+1} := \Re\aleph\Im G_i, \quad \text{for all } i, \tag{1.9.13}$$

$$\hat{G}_{i+1} := \hat{G}_i + \begin{cases} \nabla\Re\hbar G_{i+1}, & \text{for } i \text{ odd}, \\ \hbar G_{i+1}, & \text{for } i \text{ even}, \end{cases}$$

$$i := i+1,$$

until $G_i = 0$.

The resulting approximant is

$$\hat{G} = \hat{G}_{n+1} = \hbar G_1 + \nabla \Re \hbar G_2 + \hbar G_3 + \cdots . \tag{1.9.14}$$

The figure depicts the first three terms in this sum, respectively the parts between 0 and $r-1$ of the solid line, the dotted line and the dashed line.

The termination of the algorithm is at $i = n+1$ in the iteration since Nehari extension reduces the degree of the operand by one and the other operations are degree preserving.

A most noteworthy property of the Nehari Shuffle is that the upper bounds for the approximation error are available both *a priori* and *a posteriori*. These upper bounds are given by

$$\max_{\omega \in (-\pi,\pi]} \left| G(e^{j\omega}) - \hat{G}(e^{j\omega}) \right| \leq \sum_{i=1}^{n} \sigma_i(\Im G) \qquad (a\ priori), \tag{1.9.15}$$

$$\max_{\omega \in (-\pi,\pi]} \left| G(e^{j\omega}) - \hat{G}(e^{j\omega}) \right| \leq \sum_{i=1}^{n} \bar{\sigma}(\Im G_i) \qquad (a\ posteriori). \tag{1.9.16}$$

We shall explain the derivation of these bounds. Observe using (1.9.14) that

$$\begin{aligned} G - \hat{G} &= G_1 - \hbar G_1 - \nabla \Re \hbar G_2 - \hbar G_3 \cdots \\ &= G_1 - \left[G_1 - \nabla \Im G_1\right] - \nabla \Re \left[G_2 - \nabla \Im G_2\right] - \left[G_3 - \nabla \Im G_3\right] - \cdots . \end{aligned}$$

Now use (1.9.13) and the easily verified fact that $\nabla \Re \nabla = \Re$. There results

$$\begin{aligned} G - \hat{G} &= \nabla \Im G_1 - \nabla \Re \left[\Re \aleph \Im G_1 - \nabla \Im G_2\right] - \left[\Re \aleph \Im G_2 - \nabla \Im G_3\right] - \cdots \\ &= \nabla \left[\Im G_1 - \aleph \Im G_1\right] + \Re \left[\Im G_2 - \wp \Im G_2\right] + \nabla \left[\Im G_3 - \aleph \Im G_3\right] + \cdots . \end{aligned}$$

Since

$$\max_{\omega \in (-\pi,\pi]} \left| \nabla \left[\Im G_j - \aleph \Im G_j\right] \right| = \max_{\omega \in (-\pi,\pi]} \left| \left[\Im G_j - \wp \Im G_j\right] \right| = \bar{\sigma}(\Im G_j),$$

the bound (1.9.16) is immediate.

We turn now to the derivation of (1.9.15). As a first step, we establish that

$$\sigma_i(\Im G) \leq \sigma_i(G) \quad \text{for } i = 1, \ldots, n. \tag{1.9.17}$$

We know from the discussion in Section 1.6 that there exists an optimal $X_k (k = 0, \ldots, n-1)$ which is the approximation of G with k stable poles and an unspecified number of unstable poles. The precise property is

$$\min_{X_k} \max_{\omega\in(-\pi,\pi]} \left|G - X_k\right| = \sigma_{k+1}(G). \tag{1.9.18}$$

Thus, we have (with X_k subject to the pole constraint)

$$\begin{aligned}\min_{X_k} \max_{\omega\in(-\pi,\pi]} \left|\hbar G + \nabla\Im G - X_k\right| &= \min_{X_k} \max_{\omega\in(-\pi,\pi]} \left|\Im G + \nabla^{-1}\hbar G - \nabla^{-1}X_k\right| \\ &= \sigma_{k+1}(G). \end{aligned}\tag{1.9.19}$$

Consider $\nabla^{-1}\hbar G - \nabla^{-1}X_k$. The first term is an FIR system which is shifted by $r-1$ time units to be anticausal. The second term is a kth order system shifted so that the first $r-1$ terms are noncausal. Thus, the only causal and stable part of $\nabla^{-1}\hbar G - \nabla^{-1}X_k$ is $\Im X_k$, which has less than or equal to k stable poles. Next consider Y_k which is the optimal approximation of $\Im G$ with k stable poles (and an unspecified number of unstable poles). Then,

$$\min_{Y_k} \max_{\omega\in(-\pi,\pi]} \left|\Im G - Y_k\right| = \sigma_{k+1}(\Im G). \tag{1.9.20}$$

Because Y_k is optimal, we must have

$$\min_{Y_k} \max_{\omega\in(-\pi,\pi]} \left|\Im G - Y_k\right| \leq \min_{X_k} \max_{\omega\in(-\pi,\pi]} \left|\Im G + \nabla^{-1}\hbar G - \nabla^{-1}X_k\right|, \tag{1.9.21}$$

which implies (1.9.17).

From (1.9.17), (1.9.13) and (1.9.9), the following is immediate:

$$\bar{\sigma}(\Im G_2) \leq \bar{\sigma}(G_2) = \sigma_2(\Im G_1). \tag{1.9.22}$$

Similarly, we have

$$\bar{\sigma}(\Im G_3) \leq \bar{\sigma}(G_3) = \sigma_2(\Im G_2) \leq \sigma_2(G_2) = \sigma_3(\Im G_2).$$

More generally, observe the following chain of inequalities and equalities, obtained by using (1.9.17), (1.9.13) and then (1.9.9) in that order, and repeatedly:

$$\begin{aligned}\bar{\sigma}(\Im G_i) &\leq \bar{\sigma}(G_i) = \bar{\sigma}(\Re\aleph\Im G_{i-1}) = \sigma_2(\Im G_{i-1}) \\ &\leq \sigma_2(G_{i-1}) = \sigma_2(\Re\aleph\Im G_{i-2}) = \sigma_3(\Im G_{i-2}) \\ &\leq \sigma_3(G_{i-2}) \\ &\leq \sigma_{i-1}(G_2) = \sigma_{i-1}(\Re\aleph\Im G_1) = \sigma_i(\Im G). \end{aligned}\tag{1.9.23}$$

Now (1.9.15) is an immediate consequence of (1.9.16).

The *a priori* calculable upper bound provides a basis for selecting the order of the approximating FIR transfer function.

Main points of the section

1. Virtually any method for continuous-time model reduction can be used for discrete-time reduction, through incorporation of the bilinear transformation and its inverse.
2. Although truncations of balanced realizations of continuous-time systems are again balanced, this is not true in discrete time; singular perturbations of balanced realizations are again balanced in continuous time and discrete time.
3. Two methods were indicated for approximating IIR transfer functions by FIR transfer functions. One minimizes a possibly weighted L_2 norm of the additive error. The other seeks to minimize an unweighted H_∞ error norm.
4. The method aiming toward H_∞ error norm minimization, the Nehari shuffle, terminates in a finite number of steps, and upper and lower bounds on the error are readily available *a priori*.

1.10 Examples

In this section, examples are presented with a view to allowing comparison of model reduction methods.

Similarity of mode truncation and balanced realization truncation following Green and Limebeer (1995)

The system which it is sought to approximate is given by the following stable eighth order transfer function:

$$G(s) = \sum_{i=1}^{4} k_i \frac{\omega_i^2}{s^2 + 2\zeta_i \omega_i s + \omega_i^2}, \tag{1.10.1}$$

where ω_i, ζ_i, k_i can be seen in Table 1.10.1. Balanced truncation gives the following

Table 1.10.1. Resonant frequency, damping and gain of modes

i	ω_i	ζ_i	k_i
1	0.568 066 897 468 95	0.000 968 195 827 73	0.016 513 789 897 74
2	3.940 938 974 406 99	0.001 002 299 204 75	0.002 570 345 760 09
3	10.582 296 537 141 64	0.001 001 672 932 03	0.000 021 880 162 52
4	16.192 343 869 866 40	0.010 004 728 240 82	0.000 279 277 628 61

fourth order approximation:

$$G_{bt}(s) = \frac{1.326\,75 \times 10^{-6}s + 0.005\,329}{s^2 + 0.0011s + 0.3227} + \frac{1.015\,69 \times 10^{-5}s + 0.039\,920\,1}{s^2 + 0.007\,900\,02s + 15.531}. \tag{1.10.2}$$

This is almost the same as the fourth order truncated model $G_{mt}(s)$ which consists of the first and second terms of (1.10.1). Optimal Hankel norm approximation gives the following fourth order transfer function model:

$$G_{oh}(s) = \frac{3.093\,25 \times 10^{-5}s + 0.005\,327\,8}{s^2 + 0.001\,099\,76s + 0.322\,698} + \frac{0.000\,198\,16s + 0.039\,530\,3}{s^2 + 0.007\,824\,37s + 15.5304} + 1.648\,876 \times 10^{-4}. \tag{1.10.3}$$

There is a feedthrough term in $G_{oh}(s)$, but the value is small. The transfer function displays a similar frequency response to the first two reduced order models. The gain characteristics of the reduced order models are compared in Figure 1.10.1 with the fourth order model $G_{bs}(s)$ obtained via singular perturbation of the balanced realization. The similar characteristics can be seen in the pair $G_{bt}(s)$ and $G_{mt}(s)$, and in the pair $G_{oh}(s)$ and $G_{bs}(s)$. The Hankel singular values of the original system $G(s)$ are $\sigma_1 = 4.268\,193$, $\sigma_2 = 4.259\,936$, $\sigma_3 = 0.641\,754\,5$, $\sigma_4 = 0.640\,469\,3$, $\sigma_5 = 0.007\,048\,26$, $\sigma_6 = 0.006\,908\,509$, $\sigma_7 = 0.005\,465\,38$, $\sigma_8 = 0.005\,454\,559$, which means that fourth order models can be good approximants. The upper bounds of the H_∞ norm for the additive errors $G(s) - G_{bt}(s)$ and $G(s) - G_{oh}(s)$ are 0.049 753 416 and 0.024 876 708 which are calculated from (1.4.12) and (1.6.98) respectively. These values are also shown in Figure 1.10.2 with the gain plots of the additive errors.

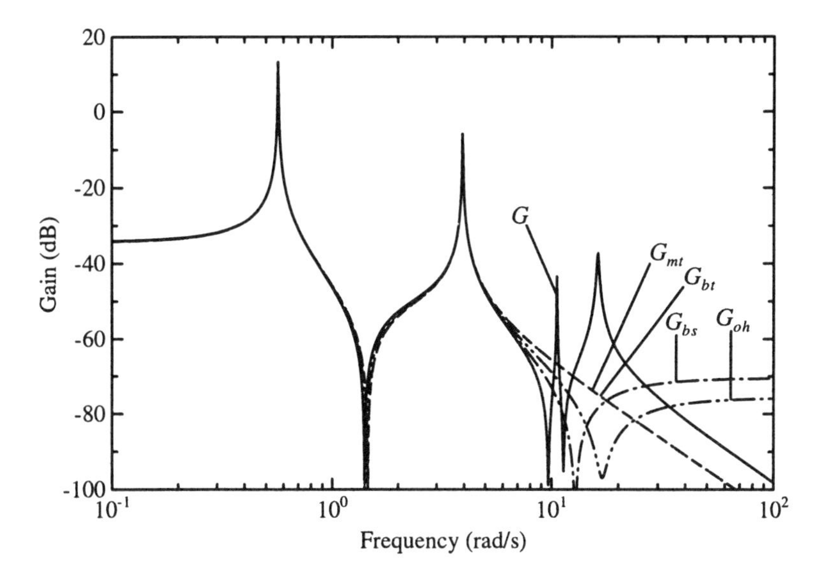

Figure 1.10.1. Comparison of the gain characteristic of the original system G and four different reduced order systems (modal truncation, balanced truncation, Hankel norm reduction, singular perturbation of a balanced realization)

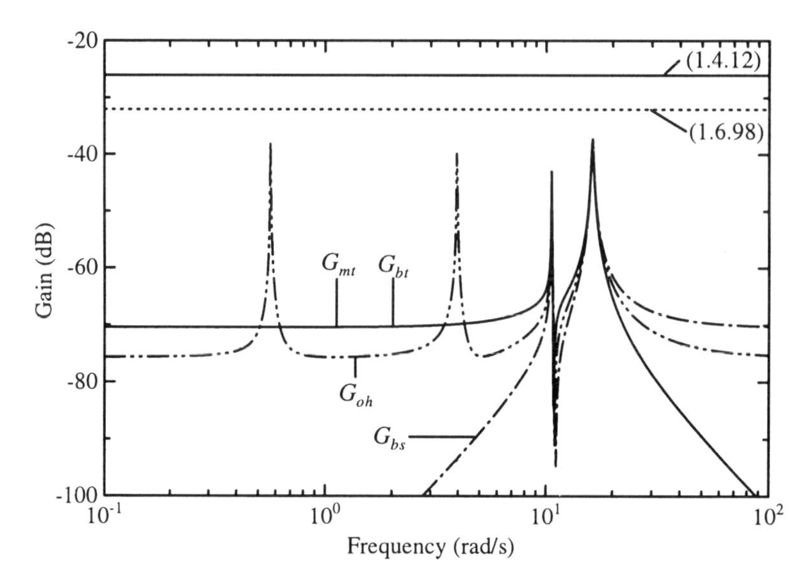

Figure 1.10.2. Additive error characteristic for four reduced order models together with error bounds

Dissimilar reduced order models for modal truncation and balanced realization truncation

In the above example, it is clear that approximate models with orders at least four will match well the original system in the low frequency region, since neglecting the high frequency modes does not have significant effect on the approximation because of the lower peak gains of the neglected modes. Another example is now given, obtained by modifying the contribution of each mode in the former example. The same values are used for (ζ_i, ω_i), $i = 1, \ldots, 4$. Set the different values for k_i as $k_1 = 0.01$, $k_2 = 0.005$, $k_3 = 0.007$, $k_4 = 0.004$. The four fourth order models (G_{bt}: balanced truncation, G_{mt}: modal truncation, G_{oh}: optimal Hankel norm approximation, G_{bs}: singular perturbation of balanced realization) are calculated and compared in the gain characteristics with the original system in Figure 1.10.3. Modal truncation tries to retain the two slower modes, while the other three methods seek to copy the first and third modes which correspond to the larger Hankel singular values. The comparison of the additive errors is shown in Figure 1.10.4. Though optimal Hankel norm approximation enjoys an attractive norm bound for the additive error, the actual error achieved is not superior to that for the other three methods in this example. Two methods of truncation give larger errors in the low frequency region; on the other hand, singular perturbation and optimal Hankel norm approximation give larger errors in the high frequency region because feedthrough terms appear in the reduced models. We have to expect larger errors for systems in which each mode has a similar contribution to the whole frequency response. We can see the similarity of contributions of modes

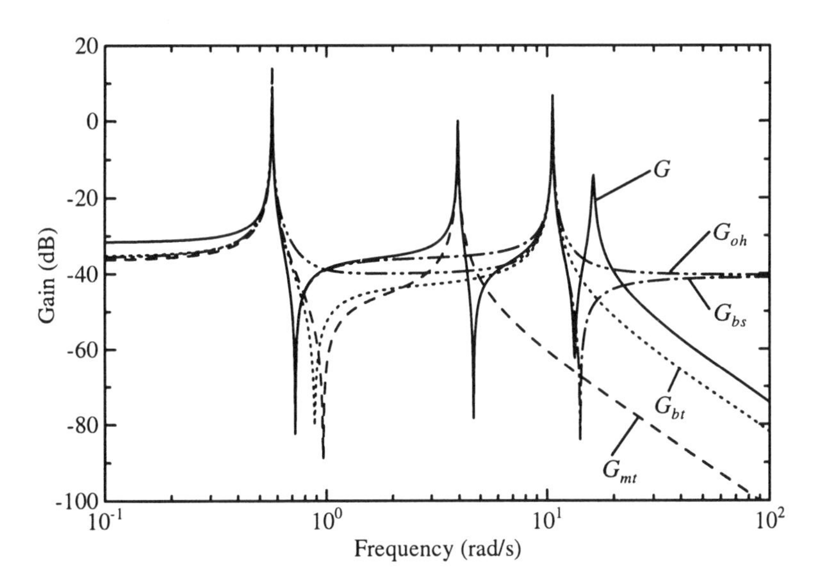

Figure 1.10.3. Comparison of the gain characteristic of the original system G and four different reduced order systems (modal truncation, balanced truncation, Hankel norm reduction, singular perturbation of a balanced realization)

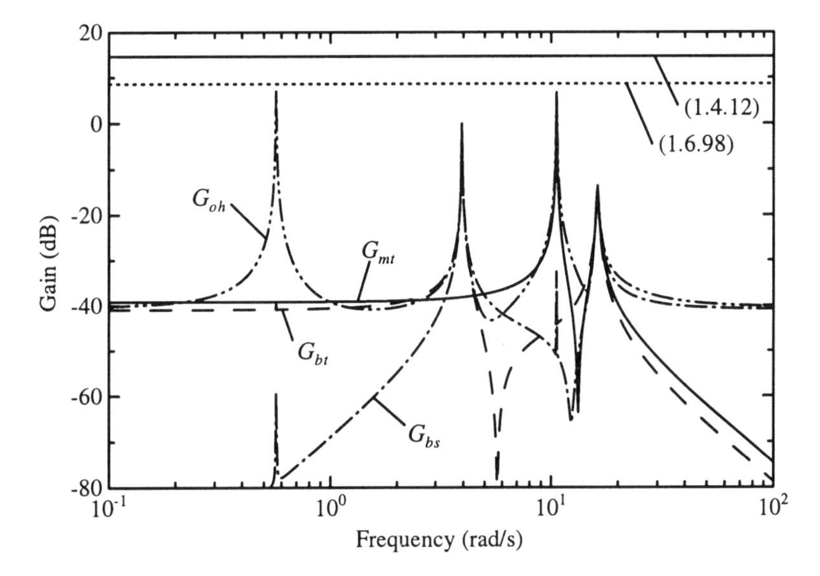

Figure 1.10.4. Additive error characteristic for four reduced order models together with error bounds

quantitatively (and also roughly) with the Hankel singular values, $\sigma_1 = 2.584\,625$, $\sigma_2 = 2.579\,625$, $\sigma_3 = 1.748\,808$, $\sigma_4 = 1.745\,308$, $\sigma_5 = 1.248\,374$, $\sigma_6 = 1.245\,874$, $\sigma_7 = 0.100\,933\,1$, $\sigma_8 = 0.098\,934$.

System with all real poles, following Liu and Anderson (1989)

Consider the following fourth order transfer function:

$$G(s) = \frac{s+4}{(s+1)(s+3)(s+5)(s+10)}. \tag{1.10.4}$$

Balanced truncation gives a better second order model than modal truncation in this example (see Figure 1.10.5). Combining H_2 norm minimization with modal truncation yields a similar reduced model to that obtained via balanced realization truncation, see the gain plots of Figure 1.10.5. (In the figure, G_{mt2} denotes the model obtained by truncation and H_2-norm optimization.) The calculation is quite easy (see (1.8.7) and (1.8.8)). The H_2 norm of the additive error decreases from $1.834\,647 \times 10^{-2}$ to $1.639\,455 \times 10^{-2}$ by the use of H_2 norm minimization on C_r.

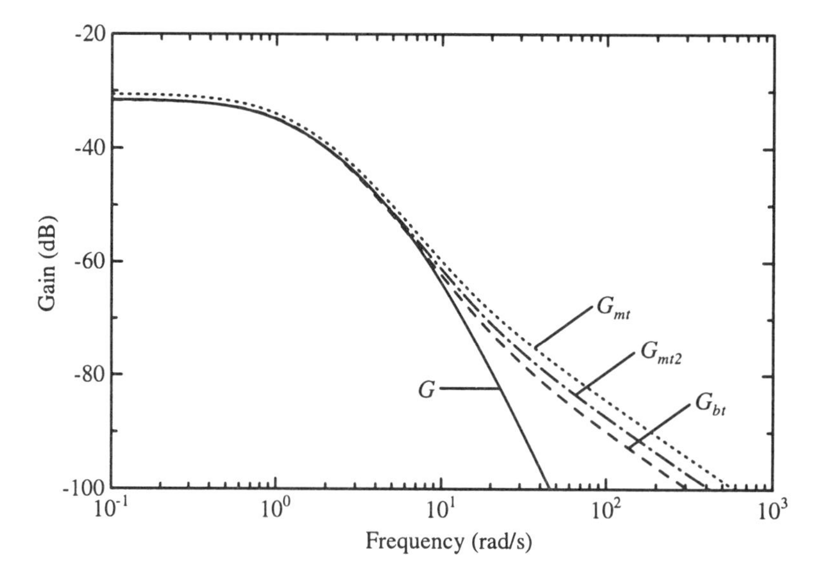

Figure 1.10.5. Comparison of the gain characteristic of the original system and three reduced order systems (balanced truncation, modal truncation and modal truncation with H_2-norm optimization)

For this example, we can also study the use of Markov parameter or time moment matching through equation error techniques. If we use the controllable canonical form for the state space realization of the original system, then the first k Markov parameters of the kth order model given by (1.5.16) with $C_r = CPL^T(LPL^T)^{-1}$ match those of

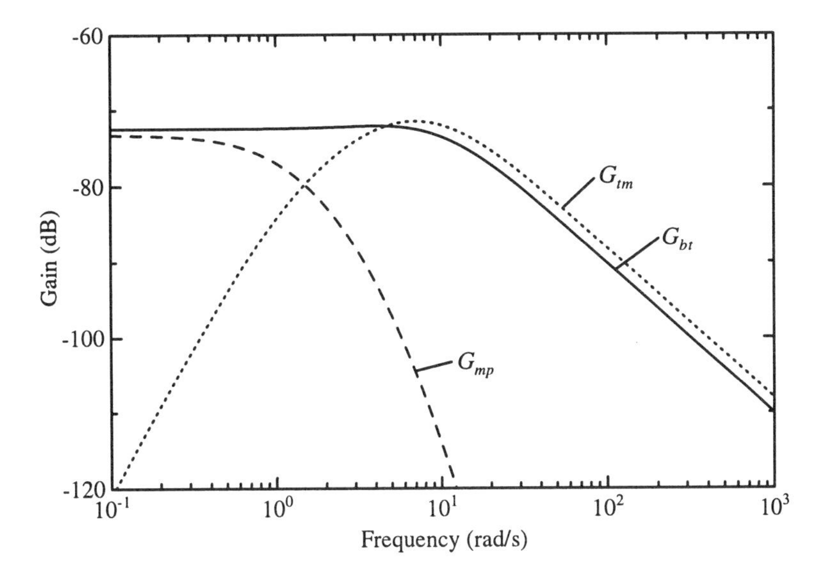

Figure 1.10.6. Comparison of gain characteristic of additive error for three reduced order models (balanced truncation, Markov parameter and time moment matching)

the original system, provided that the obtained model is controllable (Obinata, 1989). Set $k = 3$; then the calculation on (1.10.4) yields Markov parameters of the reduced model $h_0 = C_r B_r = 0$, $h_1 = C_r A_r B_r = 0$, $h_2 = C_r A_r B_r = 1$ which match those of (1.10.4). In the case of $k = 2$, the resulting reduced order model is uncontrollable. If we use the special canonical form suggested by Chidambara (1971), then the first k time moments of the kth order model which is calculated by (1.5.16) with $C_r = CPL^T(LPL^T)^{-1}$ match those of the original system. We can confirm the matching on (1.10.4) with the values: $m_0 = -C_r A_r^{-1} B_r = 2.666\,667 \times 10^{-2}$, $m_1 = C_r A_r^{-2} B_r = 3.6888 \times 10^{-2}$. The matching of Markov parameters in general gives rise to a smaller error in the high frequency region and the matching of time moments a smaller error in the low frequency region.

Chapter 2

Multiplicative Approximation

2.1 The Multiplicative Approximation Problem

In this section, we will describe what the multiplicative approximation problem is, and motivate it. In later sections we shall indicate how the problem is solved. We shall find that there is a remarkable correspondence with the earlier results on additive error approximation. There are actually two related ways of measuring the error in the multiplicative approximation problem, known as relative error and multiplicative error, and we shall consider both possibilities; the reduced order models we derive turn out to be the same.

We begin with a problem statement for the scalar transfer function case.

Approximation problem using relative error and multiplicative error

Let $G(s)$ be a transfer function of degree n. A relative error approximation of G is a transfer function $\hat{G}$ such that

$$\hat{G}(s) = G(s)\big(1 - \Delta(s)\big), \tag{2.1.1}$$

where $\Delta(s)$ is rational and stable, $\|\Delta\|_\infty < 1$. The relative error approximation problem is then to find a $\hat{G}(s)$ of lower degree than $G(s)$ such that $\|\Delta\|_\infty$ is minimized.

There is a very closely related problem, which it turns out can be solved in essentially the same way as the above relative error approximation problem; one requires

$$G(s) = \hat{G}(s)\big(1 - \Delta(s)\big) \tag{2.1.2}$$

and otherwise the problem statement is the same. This is termed a multiplicative error approximation problem.

Motivation for the problem, restrictions and general remarks

Observe that (2.1.1) implies

$$\Delta(s) = (G - \hat{G})/G. \tag{2.1.3}$$

What therefore we are seeking to minimize is a form of fractional error between G and $\hat{G}$. Such error measures arise frequently in control engineering as well as digital filter design. This is because the error measure in effect is focusing on Bode diagram errors, or errors in log-magnitudes and phases. In fact, if the errors are not large then at each frequency on the $j\omega$ axis, see Wang and Safonov (1990),

$$20\log_{10}\left|\hat{G}/G\right| \leq 8.69\,|\Delta|\,\text{dB} \quad \text{and} \quad \left|\text{phase}(G) - \text{phase}(\hat{G})\right| \leq |\Delta|\,\text{radians}.$$

There is another quite separate motivation for the multiplicative approximation problem which we shall return to in our discussion of controller reduction. It turns out that if one has an open loop model of a plant, and if one needs to simplify the model in order to design a controller—perhaps because of limitations on the controller design software—multiplicative error approximation of the plant rather than additive error approximation provides a sounder basis for simplification.

Above we have described an approximation problem for a scalar $G(s)$ with no restrictions other than that it be rational. Actually, it is frequently the case that $G(s)$ is assumed to have one or both of the following properties:

- $G(s)$ has all poles in the open left half plane.
- $G(j\omega)$ is nonzero for all ω including $\omega = \infty$.

If the restrictions are not in force the algorithms typically misbehave, and one must introduce some fix. This is less straightforward when the second constraint is violated; also, one must expect that in a great many cases, the second constraint will be violated. After all, most plants are modelled by strictly proper transfer functions.

In case $G(s)$ is multivariable, one requires preferably that $G(s)$ is square and invertible. Failing that, $G(s)$ should have full row rank. Under a full row rank assumption, the earlier statements of the approximation problems remain valid.

At the broadest level, there are two approaches to the relative/multiplicative error problem. They are in direct analogy with the two approaches of balanced truncation and Hankel norm reduction for additive error reduction. The first is known as "Balanced stochastic truncation" (abbreviated BST), and the second will be termed in this book (in order to define clearly what is involved) "multiplicative Hankel norm approximation".

Main points of the section

1. Multiplicative and relative error approximation problems can be formulated, and there are sound motivations for posing the problem.

2.2 Balanced Stochastic Truncation

The BST method appears due to Desai and Pal (1984). Clear descriptions are to be found in the work of Green (1988a), Green (1988b). See also Wang and Safonov (1990), Wang and Safonov (1991), Wang and Safonov (1992). The method, in contrast to a variant on it which introduces Hankel norms, is restricted to stable systems. If scalar, the transfer function is to be nonzero on the extended $j\omega$ axis, and if multivariable, the transfer function matrix is required to be square and nonsingular on the extended $j\omega$ axis. These latter restrictions can however be circumvented, at least to some degree.

Overview of reduction approach

Let $G(s)$ be a stable, possibly non-minimum phase, square transfer function matrix given by

$$G(s) = D + C(sI - A)^{-1}B, \tag{2.2.1}$$

where the realization is minimal.

Suppose further that $\det|G(j\omega)|$ is nonzero for all ω in $[0, \infty]$. From G one can define a power spectrum matrix $\Phi(s)$, its causal part $Z(s)$ and an additional spectral factor of $\Phi(s)$, namely $H(s)$. Without loss of generality, $H(s)$ can be taken to be stable and minimum phase, irrespective of whether $G(s)$ fulfils this property. These quantities satisfy

$$\Phi(s) = G(s)G_*(s) = H_*(s)H(s) = Z(s) + Z_*(s). \tag{2.2.2}$$

Here the notation $X_*(s)$ denotes $X^T(-s)$, *i.e.*, parahermitian conjugate. On the $j\omega$ axis, it reverts to the hermitian conjugate.

The state space description of $G(s)$ induces a state space description of $Z(s)$ of the form

$$Z(s) = D_z + C(sI - A)^{-1}L. \tag{2.2.3}$$

This state space description of $Z(s)$ then induces a state space description of $H(s)$ as

$$H(s) = D_h + K(sI - A)^{-1}L. \tag{2.2.4}$$

We can also define $F_c(s)$ using A, K and B so that

$$\begin{bmatrix} Z(s) & G(s) \\ H(s) & F_c(s) \end{bmatrix} = \begin{bmatrix} D_z & D \\ D_h & 0 \end{bmatrix} + \begin{bmatrix} C \\ K \end{bmatrix} (sI - A)^{-1} \begin{bmatrix} L & B \end{bmatrix}. \tag{2.2.5}$$

Apart from introducing precautions needed to ensure absence of numerical problems, the algorithm then proceeds by changing the coordinate basis so that the realization of $F_c(s)$ is balanced; then this realization is reduced in the usual way, and induced truncations of the realizations of $G(s)$, $H(s)$ and $Z(s)$ are simultaneously obtained. In more detail, suppose that

$$A\Sigma + \Sigma A^T + BB^T = 0, \qquad A^T\Sigma + \Sigma A + K^T K = 0, \tag{2.2.6}$$

with $\Sigma = \text{diag}[\sigma_1, \sigma_2, \ldots, \sigma_n]$ with $\sigma_1 \geq \sigma_2 \geq \cdots \geq \sigma_n > 0$. Let the *distinct* Hankel singular values of Σ be denoted by ν_i with $\nu_1 > \nu_2 > \cdots > \nu_N$.

Now Σ is partitioned into two blocks as follows:

$$\Sigma = \begin{bmatrix} \hat{\Sigma} & 0 \\ 0 & \tilde{\Sigma} \end{bmatrix}, \tag{2.2.7}$$

where $\tilde{\Sigma} = \text{diag}[\sigma_{k+1}, \ldots, \sigma_n]$ and $\sigma_k \neq \sigma_{k+1} = \nu_{r+1}$, and the other matrices are partitioned conformably as

$$\begin{bmatrix} A_{11} & A_{12} \\ A_{21} & A_{22} \end{bmatrix}, \begin{bmatrix} B_1 \\ B_2 \end{bmatrix}, \begin{bmatrix} L_1 \\ L_2 \end{bmatrix}, \begin{bmatrix} C_1 & C_2 \end{bmatrix}, \begin{bmatrix} K_1 & K_2 \end{bmatrix}.$$

Then one can form a reduced version of (2.2.5) as

$$\begin{bmatrix} \hat{Z}(s) & \hat{G}_k(s) \\ \hat{H}(s) & \hat{F}_c(s) \end{bmatrix} = \begin{bmatrix} D_z & D \\ D_h & 0 \end{bmatrix} + \begin{bmatrix} C_1 \\ K_1 \end{bmatrix} (sI - A_{11})^{-1} \begin{bmatrix} L_1 & B_1 \end{bmatrix}. \tag{2.2.8}$$

The remarkable result is the following error bound for relative error approximation, to be found in Green (1988b):

$$\sigma_{k+1} \leq \left\| G^{-1}(s)(G(s) - \hat{G}_k(s)) \right\|_\infty \leq \prod_{i=r+1}^{N} \frac{1+\nu_i}{1-\nu_i} - 1. \tag{2.2.9}$$

For the upper bound to be useful, we want $\sigma_{k+1}, \sigma_{k+2}, \ldots$ to be less than one. It turns out that all σ_i are less than or equal to one, and the number of σ_i equal to one is precisely the number of right half plane zeros of $G(s)$. The right half plane zeros of $G(s)$ are also necessarily right half plane zeros of $\hat{G}_k(s)$. For if $\hat{G}_k(s)$ did not contain a right half plane zero at the same point as each right half plane zero of $\hat{G}_k(s)$, then

$$\Delta(s) = G^{-1}(s)\big[G(s) - \hat{G}_k(s)\big] \tag{2.2.10}$$

could not be stable. As a consequence one cannot reduce $G(s)$ to a $\hat{G}_k(s)$ with degree less than the number of right half plane zeros of $G(s)$. In the practical application of the error formula, $\nu_i \leq \nu_{r+1} = \sigma_{k+1}$ is always less than one.

In the next subsections, we shall give more details concerning the algorithm (clarifying how one can avoid any potentially numerically hazardous balancing step), as well as commenting on the relevance for an alternative error criterion. In the next section, a full theoretical justification is given, and the algorithm extended to nonsquare transfer functions. We shall also prove the error bound formula.

More detail on the balanced stochastic truncation algorithm

Step 1. Determine D_z, D_h, K and L. Let P be the controllability grammian of $G(s)$, *i.e.*, the solution of the following Lyapunov equation:

$$AP + PA^T + BB^T = 0. \tag{2.2.11}$$

Then set

$$L = PC^T + BD^T. \tag{2.2.12}$$

Next, find a stabilizing solution for the following Riccati equation:

$$QA + A^T Q + (C - L^T Q)^T (DD^T)^{-1}(C - L^T Q) = 0. \tag{2.2.13}$$

(One can prove this solution always exists.)

Although D_h, D_z and K are not needed for the rest of the algorithm, we provide expressions for them as follows:

$$D_h = D^T, \qquad D_z = \left(\frac{1}{2}\right) DD^T, \qquad K = D^{-1}\left(C - L^T Q\right). \tag{2.2.14}$$

The real point of this step of the algorithm is to compute P and Q. Note that P is not only the controllability grammian of the realization of $G(s)$ but also of the realization $K(sI - A)^{-1}B$ of $F_c(s)$. Also, the definitions of K and Q ensure that

$$QA + A^T Q + K^T K = 0, \tag{2.2.15}$$

so that Q is the observability grammian of the state variable realizations of both $H(s)$ and $F_c(s)$.

Step 2. This step is a standard step for the additive error reduction of $F_c(s)$, which has controllability and observability grammians P and Q respectively obtained in Step 1. The computations were set out in the discussion of the previous chapter on balanced truncation, where we explained that using ordered Schur decompositions of PQ, we could avoid having to construct a balanced realization. The calculations lead to two nonsquare matrices L and R obeying $LR = I$.

[The matrix L here should be distinguished from that in (2.2.12)]

Step 3. The realization $\{\hat{A}, \hat{B}, \hat{C}, \hat{D}\}$ of the reduced order transfer function $\hat{G}_k(s)$ is given in terms of the realization $\{A, B, C, D\}$ of $G(s)$ and the matrices L and R found in Step 2 as follows:

$$\hat{A} = LAR, \qquad \hat{B} = LB, \qquad \hat{C} = CR, \qquad \hat{D} = D. \tag{2.2.16}$$

An alternative error criterion

In Section 2.1 of this chapter, we defined the relative error approximation problem as one of finding from a transfer function matrix G a transfer function matrix $\hat{G}$ of lower degree such that

$$\hat{G}(s) = G(s)\big(1 - \Delta(s)\big).$$

with minimization of $\|\Delta\|_\infty$, and we have described the BST algorithm for solving this problem. We commented that there is a very closely related problem, the multiplicative error approximation problem, which requires minimizing $\|\Delta\|_\infty$ where now

$$G(s) = \hat{G}(s)\big(1 - \Delta(s)\big).$$

It turns out, see Wang and Safonov (1992), that the same $\hat{G}$ can be used for the second problem, and furthermore, the same error bound applies for Δ.

Zeros on the imaginary axis, including infinity

Suppose $G(s)$ is strictly proper. Then $\det|G(j\omega)|$ has a zero at $\omega = \infty$, and the condition given earlier is violated. In this case we can use a particular bilinear transform, and set

$$\tilde{G}(s) = G\left(\frac{s}{1 - \tau s}\right), \qquad G(s) = \tilde{G}\left(\frac{s}{1 + \tau s}\right).$$

With $\tau\omega$ very small,

$$\tilde{G}(j\omega) = G\left(\frac{j\omega}{1 - \tau j\omega}\right) \simeq G(j\omega).$$

Thus approximation up to a frequency of order τ^{-1} is feasible. An approximation of order k of $\tilde{G}$ of course maps back to an approximation of order k of G.

More generally we can use $0 < a < b^{-1}$ with

$$\tilde{G}(s) = G\left(\frac{s - a}{-bs + 1}\right), \qquad G(s) = \tilde{G}\left(\frac{s + a}{bs + 1}\right).$$

In this case, the values of G along the $j\omega$ axis are the same as the values of $\tilde{G}(s)$ on a circle in $\mathrm{Re}[s] > 0$ with diameter $[a, b^{-1}]$, while the values of $\tilde{G}$ along the $j\omega$-axis are the same as the values of $G(s)$ around a circle with diameter $[-b^{-1} + j0, -a + j0]$

Zeros of $G(s)$ on the $j\omega$-axis, including ∞, as well as zeros in $\mathrm{Re}[s] > 0$ all become zeros in $\mathrm{Re}[s] > 0$ of $\tilde{G}(s)$. If $G(s)$ has no zero on the circle with diameter $[-b^{-1} + j0, -a + j0]$, $\tilde{G}(s)$ will be nonsingular on the extended imaginary axis, and the BST reduction algorithm can be applied to $\tilde{G}$.

If a, b are small enough, $G(s)$ will have all poles inside the circle with diameter $[-b^{-1} + j0, -a + j0]$ and $\tilde{G}(s)$ will have all poles in $\mathrm{Re}[s] < 0$.

Main points of the section

1. An algorithm requiring the solution of a Lyapunov equation and a Riccati equation is available for relative error reduction; balanced truncation is part of the algorithm.

2. An error bound is available, involving Hankel singular values.

3. The same procedure and bound applies for multiplicative error reduction.

4. Zeros on the imaginary axis can be accommodated using a bilinear transformation, reduction, and then inverse bilinear transformation.

2.3 Balanced Stochastic Truncation: Theory and Extensions

In this section, we shall present a number of results (with proofs) which justify the various assertions made in the previous section regarding the BST algorithm. In the process, we shall also note some extensions to the algorithms—to the nonsquare case especially.

Construction of $Z(s)$, $H(s)$ and $F_c(s)$

We begin with the construction of $Z(s)$, $H(s)$ and $F_c(s)$ from a minimal realization $\{A, B, C, D\}$.

Lemma 2.3.1. *Let $G(s)$ be a stable square transfer function matrix with minimal realization*

$$G(s) = D + C(sI - A)^{-1}B. \tag{2.3.1}$$

Let P solve the Lyapunov equation

$$AP + PA^T + BB^T = 0 \tag{2.3.2}$$

and let

$$L = PC^T + BD^T \tag{2.3.3}$$

and

$$Z(s) = \frac{1}{2}DD^T + C(sI - A)^{-1}L. \tag{2.3.4}$$

Then

$$Z(s) + Z^T(-s) = G(s)G^T(-s). \tag{2.3.5}$$

Proof. Observe that

$$G(s)G^T(-s) = DD^T + DB^T(-sI - A^T)^{-1}C^T + C(sI - A)^{-1}BD^T \\ +C(sI - A)^{-1}BB^T(-sI - A^T)^{-1}C^T.$$

Since

$$\begin{aligned}(sI - A)^{-1}BB^T(-sI - A^T)^{-1} &= (sI - A)^{-1}\left[(sI - A)P + P(-sI - A^T)\right] \\ &\quad \times(-sI - A^T)^{-1} \\ &= P(-sI - A^T)^{-1} + (sI - A)^{-1}P\end{aligned}$$

there holds

$$G(s)G^T(-s) = DD^T + (DB^T + CP)(-sI - A^T)^{-1}C^T \\ +C(sI - A)^{-1}(BD^T + P^TC),$$

as required. □

The construction of $H(s)$ is almost as straightforward.

Lemma 2.3.2. *With the same hypothesis as Lemma 2.3.1, and with $G(j\omega)$ assumed to be nonsingular on the extended imaginary axis there exists a stabilizing solution Q of*

$$XA + A^TX + (C - L^TX)^T(DD^T)^{-1}(C - L^TX) = 0, \tag{2.3.6}$$

which is positive definite and for which $X - Q \geq 0$ for all the solutions X. Moreover, the definition

$$K = D^{-1}(C - L^TQ) \tag{2.3.7}$$

ensures that

$$H(s) = D^T + K(sI - A)^{-1}L \tag{2.3.8}$$

obeys

$$Z(s) + Z^T(-s) = H^T(-s)H(s), \tag{2.3.9}$$

with $H(s)$ minimum phase.

Proof. The fact that (2.3.6) has a stabilizing solution follows from the state-variable realization of $Z(s)$, and the fact that

$$Z(j\omega) + Z^T(-j\omega) = G(j\omega)G^T(-j\omega) > 0, \quad \text{for all real } \omega. \tag{2.3.10}$$

(Recall that $G(j\omega)$ is assumed nonsingular for all real ω). We are appealing here to fairly standard results on steady-state Riccati equations, see *e.g.*, Green and Limebeer

(1995). An argument like that of Lemma 2.3.1 then establishes (2.3.9). The zeros of $H(s)$ are the eigenvalues of

$$A - LD^{-T}K = A - L(DD^T)^{-1}(C - L^T Q) \tag{2.3.11}$$

which lie in $\text{Re}[s] < 0$ because of the stabilizing property of Q. □

The transfer function $F_c(s)$ is calculated in a very straightforward way. It has some key properties.

Lemma 2.3.3. *Adopt the hypotheses of Lemma 2.3.1 and Lemma 2.3.2. Define $F_c(s)$ by*

$$F_c(s) = K(sI - A)^{-1}B. \tag{2.3.12}$$

Then

$$F_c(s) = \left[H_*^{-1}(s)G(s)\right]_+ - H_*^{-1}(\infty)G(\infty). \tag{2.3.13}$$

[The notation $[X(s)]_+$ denotes the sum of the strictly proper partial fraction summands with poles in $\text{Re}[s] < 0$*, plus $X(\infty)$.] Moreover, the Hankel singular values of $F_c(s)$ are bounded by* 1.

Proof. Observe that

$$\begin{aligned}
H_*(s)F_c(s) &= \left[D + L^T\left(-sI - A^T\right)^{-1}K^T\right]K\left(sI - A\right)^{-1}B \\
&= DK\left(sI - A\right)^{-1}B \\
&\quad + L^T\left(-sI - A^T\right)^{-1}\left[\left(-sI - A^T\right)Q + Q\left(sI - A\right)\right]\left(sI - A\right)^{-1}B \\
&= \left(DK + L^T Q\right)\left(sI - A\right)^{-1}B + L^T\left(-sI - A^T\right)^{-1}QB \\
&= G(s) + \left[-D + L^T\left(-sI - A^T\right)^{-1}QB\right].
\end{aligned}$$

Since $H(s)$ is minimum phase, $H_*^{-1}(s)$ has all its poles in $\text{Re}[s] \geq 0$. Multiplying the above equation by $H_*^{-1}(s)$ on the left, it is immediate that

$$F_c(s) = \left[H_*^{-1}(s)G(s)\right]_+ - H_*^{-1}(\infty)G(\infty).$$

We shall give two proofs of the Hankel singular value claim. First, observe that $H_*^{-1}(s)G(s)$ is an all-pass (although not normally stable), since $G(s)G_*(s) = H_*(s)H(s)$. Hence there exists an unstable $X(s)$ for which

$$\left\| F_c(s) + X(s) \right\|_\infty = 1.$$

This means that the largest Hankel singular value of $F_c(s)$ is at most 1. For the alternative argument, observe from (2.3.6) that Q^{-1} satisfies

$$AQ^{-1} + Q^{-1}A^T + \left(CQ^{-1} - L^T\right)^T\left(DD^T\right)\left(CQ^{-1} - L^T\right) = 0 \tag{2.3.14}$$

and is the maximum solution of the equation

$$AY + YA^T + (YC^T - L^T)(DD^T)^{-1}(CY - L) = 0. \tag{2.3.15}$$

As the maximum solution, it is antistabilizing. Moreover, from (2.3.2) and (2.3.3), P satisfies

$$AP + PA^T + (PC^T - L)(DD^T)^{-1}(CP - L^T) = 0. \tag{2.3.16}$$

Hence $Q^{-1} - P \geq 0$. It is immediate that $\lambda_{\max}(QP) \leq 1$.

Now P and Q are respectively the controllability and observability grammians for the realization $\{A, B, K\}$ of $F_c(s)$. Hence $\sigma_1^2[F_c(s)] = \lambda_{\max}(QP) \leq 1$. □

Zeros of $G(s)$, the stability of $\Delta(s)$ and simplification for minimum phase $G(s)$

Suppose now that (through coordinate basis change if necessary)

$$P = Q = \begin{bmatrix} I & 0 \\ 0 & \bar{\Sigma} \end{bmatrix}, \tag{2.3.17}$$

where $I - \bar{\Sigma}$ is nonsingular, and $\bar{\Sigma}$ is diagonal with decreasing entries. Suppose that A, B, C are partitioned conformably:

$$A = \begin{bmatrix} A_{11} & A_{12} \\ A_{21} & A_{22} \end{bmatrix}, \qquad B = \begin{bmatrix} B_1 \\ B_2 \end{bmatrix}, \qquad C = \begin{bmatrix} C_1 & C_2 \end{bmatrix}. \tag{2.3.18}$$

These definitions allow us to identify the zeros of $G(s)$.

Lemma 2.3.4. *Adopt the hypotheses of Lemma 2.3.1 and Lemma 2.3.2 and suppose the coordinate basis is chosen so that P and Q are balanced, as in* (2.3.17), *with $0 < \bar{\Sigma} < I$. Suppose A, B, C are partitioned as in* (2.3.18). *Then*

$$A - BD^{-1}C = \begin{bmatrix} A_{11} - B_1 D^{-1} C_1 & 0 \\ * & A_{22} - B_2 D^{-1} C_2 \end{bmatrix}. \tag{2.3.19}$$

and the unstable zeros of $G(s)$ are eigenvalues of $A_{11} - B_1 D^{-1} C_1$, and the stable zeros are eigenvalues of $A_{22} - B_2 D^{-1} C_2$, so that the number of unstable zeros of $G(s)$ is the number of Hankel singular values of $F_c(s)$ equal to 1.

Proof. From (2.3.14) and (2.3.16) there follows

$$\begin{aligned} (Q^{-1} - P)\left[A^T - C^T(DD^T)^{-1}L^T + C^T(DD^T)^{-1}CP\right] \\ + \left[A - L(DD^T)^{-1}C + PC^T(DD^T)^{-1}C\right](Q^{-1} - P) \\ + (Q^{-1} - P)C^T(DD^T)^{-1}C(Q^{-1} - P) = 0. \end{aligned} \tag{2.3.20}$$

Suppose that

$$A^T - C^T(DD^T)^{-1}L^T + C^T(DD^T)^{-1}CP = E = \begin{bmatrix} E_{11} & E_{12} \\ E_{21} & E_{22}. \end{bmatrix} \tag{2.3.21}$$

Notice that the matrix on the left is $A^T - C^T D^{-T} B^T$, and its eigenvalues are therefore the zeros of $G(s)$. Since

$$Q^{-1} - P = \begin{bmatrix} 0 & 0 \\ 0 & \bar{\Sigma}^{-1} - \bar{\Sigma} \end{bmatrix}, \tag{2.3.22}$$

it easily follows from (2.3.20) that $(\bar{\Sigma}^{-1} - \bar{\Sigma})E_{21} = 0$ or that $E_{21} = 0$, corresponding to the zero 1-2 block in (2.3.19). Also,

$$(\bar{\Sigma}^{-1} - \bar{\Sigma})E_{22} + E_{22}^T(\bar{\Sigma}^{-1} - \bar{\Sigma}) + (\bar{\Sigma}^{-1} - \bar{\Sigma})C_2^T(DD^T)^{-1}C_2(\bar{\Sigma}^{-1} - \bar{\Sigma}) = 0. \tag{2.3.23}$$

It follows that $\operatorname{Re} \lambda_i(E_{22}) \leq 0$. Since we proved that $E_{21} = 0$, we know that

$$\{\text{zeros of } G(s)\} = \{\lambda_i(E_{11})\} \cup \{\lambda_i(E_{22})\}.$$

Since $G(s)$ has no purely imaginary zeros, $\operatorname{Re} \lambda_i(E_{22}) < 0$.

Next, because Q^{-1} is antistabilizing, the following matrix has all eigenvalues in $\operatorname{Re}[s] > 0$:

$$\begin{aligned} A^T &- C^T(DD^T)^{-1}(L^T - CQ^{-1}) \\ &= A^T - C^T(DD^T)^{-1}(L^T - CP) + C^T(DD^T)^{-1}C(Q^{-1} - P) \\ &= \begin{bmatrix} E_{11} & E_{12} \\ 0 & E_{22} \end{bmatrix} + \begin{bmatrix} 0 & * \\ 0 & * \end{bmatrix} = \begin{bmatrix} E_{11} & * \\ 0 & * \end{bmatrix}. \end{aligned}$$

Hence $\operatorname{Re} \lambda_i(E_{11}) > 0$. Since (2.3.19) is the transpose of (2.3.21), the claim of the Lemma is established. □

In the truncation operation that leads to reduced order $\hat{G}(s)$, the matrices A_{11}, B_1 and C_1 are left untouched, while the matrices A_{22}, B_2 and C_2 are shrunk as part of the truncation. Suppose that

$$\begin{bmatrix} I & 0 \\ 0 & \bar{\Sigma} \end{bmatrix} = \begin{bmatrix} I & 0 & 0 \\ 0 & \hat{\Sigma} & 0 \\ 0 & 0 & \tilde{\Sigma} \end{bmatrix}.$$

(The notation is a little different to earlier). With $\hat{E}_{22} = (\hat{A}_{22} - \hat{B}_2 D^{-1}\hat{C}_2)^T$, (2.3.23) holds with hats introduced, and it is evident that

$$\{\text{zeros of } \hat{G}(s)\} = \{\lambda_i(E_{11})\} \cup \{\lambda_i(\hat{E}_{22})\}.$$

This shows that the unstable zeros of $G(s)$ are copied to $\hat{G}(s)$. In fact, it can be shown that $G^{-1}(s)\hat{G}(s)$ is stable. Therefore our earlier claim that

$$G^{-1}(G - \hat{G}) = \Delta,$$

for some stable Δ is validated.

As an addendum, let us discuss what happens when $G(s)$ is minimum phase. Then $(Q^{-1} - P)$ is nonsingular, and for (2.3.20), we see that $R = (Q^{-1} - P)^{-1}$ satisfies

$$R\left(A - BD^{-1}C\right) + \left(A - BD^{-1}C\right)^T R + C^T\left(DD^T\right)^{-1}C = 0,$$

so that R is the observability grammian of $G^{-1}(s)$. If P, the controllability grammian of $G(s)$, is balanced against R, the observability grammian of $G^{-1}(s)$, so that

$$P = R = \mathrm{diag}(\alpha_1, \alpha_2, \ldots, \alpha_n),$$

with $\alpha_i > \alpha_{i+1}$, then Q is diagonal and

$$\sigma_i = \frac{\alpha_i}{\sqrt{1+\alpha_i^2}}.$$

No Riccati equation needs to be solved in this case!

Extension of algorithm to nonsquare transfer function matrices

Suppose now that $G(s)$ is not square, but that $\Phi(s) = G(s)G_*(s)$ is nonsingular on the $j\omega$-axis. This means that $G(s)$ is fat (more columns than rows), with $D = G(s)$ of full row rank. It is evident that a square minimum phase $H(s)$ still exists so that $\Phi(s) = H_*(s)H(s)$, and this suggests that perhaps the same algorithm can be used.

An alternative approach is based on augmentation. Suppose that we can find $\tilde{G}(s)$ such that

$$G_a(s) = \begin{bmatrix} G(s) \\ \tilde{G}(s) \end{bmatrix}, \tag{2.3.24}$$

with $G_a(s)$ square and with the same Hankel singular values for $F_{ac}(s)$ as for $F_c(s)$. Then we can reduce to obtain

$$\begin{bmatrix} \hat{G}(s) \\ \hat{\tilde{G}}(s) \end{bmatrix} = \begin{bmatrix} G(s) \\ \tilde{G}(s) \end{bmatrix}\left[I - \Delta_a(s)\right]. \tag{2.3.25}$$

The matrices $\tilde{G}(s)$ and $\hat{\tilde{G}}(s)$ are now irrelevant and

$$\hat{G}(s) = G(s)\left[I - \Delta_a(s)\right]. \tag{2.3.26}$$

The calculation is done as follows. Let $\tilde{D}$ be a matrix for which $[D^T \ \ \tilde{D}^T]$ has full rank and which has columns orthogonal to those of D, so that $D\tilde{D}^T = 0$. With P and Q as before, select $\tilde{L}$ to satisfy

$$(I - PQ)\tilde{L} = B\tilde{D}^T. \tag{2.3.27}$$

(That this is possible will be demonstrated below. Since $I - PQ$ may be singular, this is a nontrivial point). Last, set

$$\tilde{C} = \tilde{L}^T Q, \tag{2.3.28}$$

so that

$$G_a(s) = \left[\begin{array}{c|c} A & B \\ \hline \begin{bmatrix} C \\ \tilde{C} \end{bmatrix} & \begin{matrix} D \\ \tilde{D} \end{matrix} \end{array}\right]. \tag{2.3.29}$$

Observe that D_a (with obvious definition) is square and nonsingular. The controllability grammian of $G_a(s)$ is the same as that of $G(s)$, *viz.* P.

To follow the algorithm now for $G_a(s)$, we next set

$$\begin{aligned} L_a &= PC_a^T + BD_a^T \\ &= \begin{bmatrix} PC^T + BD^T & P\tilde{C}^T + B\tilde{D}^T \end{bmatrix} \\ &= \begin{bmatrix} L & PQ\tilde{L} + (I - PQ)\tilde{L} \end{bmatrix} \\ &= \begin{bmatrix} L & \tilde{L} \end{bmatrix}. \end{aligned} \tag{2.3.30}$$

Next, let Q_a denote the matrix derived for $G_a(s)$ in the same manner as Q is derived for G. Thus Q_a is the unique stabilizing solution of

$$Q_a A + A^T Q_a + [C_a - L_a^T Q_a]^T (D_a D_a^T)^{-1} [C_a - L_a^T Q_a] = 0$$

or

$$Q_a A + A^T Q_a$$

$$+ \begin{bmatrix} C^T - Q_a L & \tilde{C}^T - Q_a \tilde{L} \end{bmatrix} \begin{bmatrix} (DD^T)^{-1} & 0 \\ 0 & (\tilde{D}\tilde{D}^T)^{-1} \end{bmatrix} \begin{bmatrix} C - L^T Q_a \\ \tilde{C} - \tilde{L}^T Q_a \end{bmatrix} = 0.$$

The choice $Q_a = Q$ leads to a solution, by virtue of (2.3.28) and the underlying equation for Q, and it is of course stabilizing. Therefore, by uniqueness, it is the solution being sought. Consequently, the two grammians for $F_{ca}(s)$ are the same as those for $F_c(s)$, *viz.* P and Q, and the same error formula applies. Indeed, we will obtain the same reduced order $\hat{G}_k(s)$ using the augmentation approach as we do by proceeding directly. What we obtain as a bonus is proof of the error bound, and the formula (2.3.26) for a stable $\Delta_a(s)$.

It remains to demonstrate the solvability of (2.3.27) for $\tilde{L}$. This is immediate from the following Lemma.

Lemma 2.3.5. *Adopt the hypotheses of Lemma 2.3.1 and Lemma 2.3.2 save that $G(s)$ has more columns than rows and has full rank on the extended $j\omega$-axis. Let U span the nullspace of $Q^{-1} - P$. Then $\tilde{D}B^T U = 0$.*

Proof. Recall that $AP + PA^T + BB^T = 0$ and $L = PC^T + BD^T$ (see Lemma 2.3.1). Hence

$$B\begin{pmatrix} D^T & \tilde{D}^T \end{pmatrix} = \begin{bmatrix} L - PC & B\tilde{D}^T \end{bmatrix}$$

and

$$\begin{aligned} BB^T &= \begin{bmatrix} L - PC & B\tilde{D}^T \end{bmatrix} (D_a D_a^T)^{-1} \begin{bmatrix} (L - PC)^T \\ \tilde{D}B^T \end{bmatrix} \\ &= (L - PC)(DD^T)^{-1}(L - PC)^T + B\tilde{D}^T(\tilde{D}\tilde{D}^T)^{-1}\tilde{D}B^T \end{aligned}$$

or

$$PA + A^T P + (L - PC)(DD^T)^{-1}(L - PC)^T + B\tilde{D}^T(\tilde{D}\tilde{D}^T)^{-1}\tilde{D}B^T = 0. \tag{2.3.31}$$

Equation (2.3.31) is a replacement for (2.3.16). The corresponding replacement for (2.3.20) is

$$\begin{aligned} &(Q^{-1} - P)\left[A^T - C^T(DD^T)^{-1}L^T + C^T(DD^T)^{-1}CP\right] \\ &\quad + \left[A - L(DD^T)^{-1}C + PC^T(DD^T)^{-1}C\right](Q^{-1} - P) \\ &\quad + (Q^{-1} - P)C^T(DD^T)^{-1}C(Q^{-1} - P) + B\tilde{D}^T(\tilde{D}\tilde{D}^T)^{-1}\tilde{D}B^T = 0. \end{aligned}$$

Pre- and post-multiplying by U^T and U implies that

$$\tilde{D}B^T U = 0.$$

□

Establishing the lower error bound

In this subsection, we shall establish the lower bound of (2.2.9). In fact we shall show that for any $X_k \in RH_\infty^-(k)$

$$\sigma_{k+1} \leq \left\| G^{-1}(s)\big(G(s) - X_k(s)\big) \right\|_\infty. \tag{2.3.32}$$

Then (2.2.9) becomes a special case. The argument is as follows. With $F = H_*^{-1}G$, and the σ_i the Hankel singular values of the causal part of F, recall that

$$\inf_{\hat{F}_k \in RH_\infty^-(k)} \left\| F - \hat{F}_k \right\|_\infty = \sigma_{k+1}.$$

Now because H_*^{-1} has all poles and zeros in $\mathrm{Re}[s] > 0$,

$$\begin{aligned}
\sigma_{k+1} &= \inf_{\hat{F}_k \in RH_\infty^-(k)} \left\| F - \hat{F}_k \right\|_\infty \\
&= \inf_{X_k \in RH_\infty^-(k)} \left\| H_*^{-1}G - H_*^{-1}X_k \right\|_\infty \\
&= \inf_{X_k \in RH_\infty^-(k)} \left\| H_*^{-1}(G - X_k) \right\|_\infty \\
&= \inf_{X_k \in RH_\infty^-(k)} \left\| G^{-1}(G - X_k) \right\|_\infty .
\end{aligned}$$

Establishing the upper error bound

In this subsection, we shall establish the upper error bound of (2.2.9). The technique of Green (1988b) will not be used because it depends on a knowledge of the error bound for multiplicative Hankel approximation. Instead, we shall use some arguments from Wang and Safonov (1991).

We begin with more preliminary calculations and observations. Let $\hat{G}(s)$ for the moment denote a reduced order approximation of $G(s)$ of indeterminate order. Assuming $[A, B, K]$ is balanced and that (2.2.6) to (2.2.8) hold, define

$$\tilde{A}_g(s) = A_{22} + A_{21}(sI - A_{11})^{-1}A_{12}, \tag{2.3.33}$$

$$\tilde{B}_g(s) = B_2 + A_{21}(sI - A_{11})^{-1}B_1, \tag{2.3.34}$$

$$\tilde{C}_g(s) = C_2 + C_1(sI - A_{11})^{-1}A_{12}. \tag{2.3.35}$$

Then it is not difficult to check that

$$G(s) - \hat{G}(s) = \tilde{C}_g(s)\left[sI - \tilde{A}_g(s)\right]^{-1}\tilde{B}_g(s) \tag{2.3.36}$$

and

$$\tilde{\Sigma}\tilde{A}_{g*}(s) + A_g(s)\tilde{\Sigma} + \tilde{B}_g(s)\tilde{B}_{g*}(s) = 0. \tag{2.3.37}$$

In fact, with also

$$\tilde{L}_g(s) = L_2 + A_{21}(sI - A_{11})^{-1}L_1, \tag{2.3.38}$$

$$\tilde{K}_g(s) = K_2 + K_1(sI - A_{11})^{-1}A_{12}, \tag{2.3.39}$$

there holds

$$\begin{bmatrix} Z(s) & G(s) \\ H(s) & F_c(s) \end{bmatrix} = \begin{bmatrix} \hat{Z}(s) & \hat{G}(s) \\ \hat{H}(s) & \hat{F}_c(s) \end{bmatrix} + \begin{bmatrix} \tilde{C}_g(s) \\ \tilde{K}_g(s) \end{bmatrix} \left[sI - \tilde{A}_g(s)\right]^{-1} \begin{bmatrix} \tilde{L}_g(s) & \tilde{B}_g(s) \end{bmatrix} \tag{2.3.40}$$

and

$$\tilde{A}_{g*}(s)\tilde{\Sigma} + \tilde{\Sigma}\tilde{A}_g(s) + \tilde{K}_{g*}(s)\tilde{K}_g(s) = 0. \tag{2.3.41}$$

From these equations, it follows with simple calculations that $\hat{F}_c$, $\hat{G}_c$, $\hat{H}_c$ and $\hat{Z}$ are related in the same manner as F_c, G, H and Z, *i.e.*,

$$\hat{G}(s)\hat{G}_*(s) \equiv \hat{H}_*(s)\hat{H}(s) = \hat{Z}(s) + \hat{Z}_*(s), \tag{2.3.42}$$

$$\hat{H}(s) \text{ is minimum phase}, \tag{2.3.43}$$

$$\hat{F}_c(s) = \left[\hat{H}_*^{-1}(s)\hat{G}(s)\right]_+ - \hat{H}_*^{-1}(\infty)\hat{G}(\infty). \tag{2.3.44}$$

We claim also that

$$\hat{H}_*(s)\tilde{K}_g(s) + \tilde{L}_{g*}(s)\tilde{\Sigma} = \tilde{C}_g(s) \tag{2.3.45}$$

and

$$\tilde{B}_g(s)\hat{G}_*(s) + \tilde{\Sigma}\tilde{C}_{g*}(s) = \tilde{L}_g(s). \tag{2.3.46}$$

We will establish (2.3.45) only. Observe that

$$\begin{aligned} \hat{H}_*(s)\tilde{K}_g(s) =& \left[D + L_1^T(-sI - A_{11}^T)^{-1}K_1^T\right]\left[K_2 + K_1(sI - A_{11})^{-1}A_{12}\right] \\ =& \, DK_2 + DK_1(sI - A_{11})^{-1}A_{12} + L_1^T(-sI - A_{11}^T)^{-1}K_1^T K_2 \\ &+ L_1^T(-sI - A_{11}^T)^{-1}K_1^T K_1(sI - A_{11})^{-1}A_{12} \\ =& \, DK_2 + (DK_1 + L_1^T\hat{\Sigma})(sI - A_{11})^{-1}A_{12} \\ &+ L_1^T(-sI - A_{11}^T)\left[K_1^T K_2 + \hat{\Sigma}A_{12}\right], \end{aligned}$$

[on using $K_1^T K_1 = (-sI - A_{11}^T)\hat{\Sigma} + \hat{\Sigma}(sI - A_{11})$]. Also, $\tilde{L}_{g*}(s)\tilde{\Sigma} = L_2^T\tilde{\Sigma} + L_1^T(-sI - A_{11}^T)^{-1}A_{21}^T\tilde{\Sigma}$.

From $\Sigma A + A^T\Sigma + K^T K = 0$ it follows that $K_1^T K_2 + \hat{\Sigma}A_{12} + A_{21}^T\tilde{\Sigma} = 0$. Adding the expressions for $\hat{H}_*(s)\tilde{K}_g(s)$ and $\tilde{L}_{g*}(s)\tilde{\Sigma}_2$, and using the fact $C = L^T\Sigma + DK$ yields (2.3.45).

Next we have

Lemma 2.3.6. *With notation as above, the error functions* $F_c(s) - \tilde{F}_c(s)$, $G(s) - \tilde{G}(s)$ *and* $H(s) - \tilde{H}(s)$ *are related by*

$$H_*(s)\left[F_c(s) - \hat{F}_c(s)\right] = G(s) - \hat{G}(s) + \tilde{L}_{g*}(s)\left[-sI - \tilde{A}_{g*}(s)\right]^{-1}\tilde{\Sigma}\tilde{B}_g(s), \tag{2.3.47}$$

$$\left[F_c(s) - \hat{F}_c(s)\right]G_*(s) = H(s) - \hat{H}(s) + \tilde{K}_g(s)\tilde{\Sigma}\left[-sI - \tilde{A}_{g*}(s)\right]^{-1}\tilde{C}_{g*}(s). \tag{2.3.48}$$

Proof. By (2.3.40), there holds

$$
\begin{aligned}
H_*(s)\big[F_c(s) - \hat{F}_c(s)\big] \\
&= \hat{H}_*(s)\tilde{K}_g(s)\big[sI - \tilde{A}_g(s)\big]^{-1}\tilde{B}_g(s) \\
&\quad + \tilde{L}_{g*}(s)\big[-sI - \tilde{A}_{g*}(s)\big]^{-1}\tilde{K}_{g*}(s)\tilde{K}_g(s)\big[sI - \tilde{A}_g(s)\big]^{-1}\tilde{B}_g(s).
\end{aligned}
$$

Using (2.3.41) results in

$$
\begin{aligned}
H_*(s)\big[F_c(s) - \hat{F}_c(s)\big] \\
&= \hat{H}_*(s)\tilde{K}_g(s)\big[sI - \tilde{A}_g(s)\big]^{-1}\tilde{B}_g(s) \\
&\quad + \tilde{L}_{g*}(s)\tilde{\Sigma}\big[sI - \tilde{A}_g(s)\big]\tilde{B}_g(s) + \tilde{L}_{g*}(s)\big[-sI - \tilde{A}_{g*}(s)\big]^{-1}\tilde{\Sigma}\tilde{B}_g(s).
\end{aligned}
$$

Using (2.3.45), we obtain (2.3.47). Equation (2.3.48) is established in a similar way. □

With these long preliminary calculations out of the way, we can now make progress on calculating the error bound itself. Perhaps unsurprisingly, we consider first a reduction of $G(s)$ to $\hat{G}(s)$ where $\tilde{\Sigma} = \sigma_n I$, *i.e.*, we reduce by an amount equal to the order of the smallest Hankel singular value of $F_c(s)$.

Lemma 2.3.7. *With notation as above, and with the additional assumption that $\tilde{\Sigma} = \sigma_n I$, there holds*

$$
\left\| G^{-1}(s)\big[G(s) - \hat{G}(s)\big] \right\|_\infty = \left\| H_*^{-1}(s)\big[G(s) - \hat{G}(s)\big] \right\|_\infty \le \frac{2\sigma_n}{1-\sigma_n}. \tag{2.3.49}
$$

Proof. The first equality in (2.3.49) is a trivial consequence of $GG_* = H_*H$. For then $G^{-1}H_*$ is all-pass and

$$
G^{-1}(G - \hat{G}) = (G^{-1}H_*)\big[H_*^{-1}(G - \hat{G})\big].
$$

Now from (2.3.47), we have

$$
\begin{aligned}
\left\| H_*^{-1}(G - \hat{G}) \right\|_\infty &= \left\| F_c - \hat{F}_c - H_*^{-1}\tilde{L}_{g*}(-sI - \tilde{A}_{g*})^{-1}\tilde{\Sigma}\tilde{B}_g \right\|_\infty \\
&\le 2\sigma_n + \sigma_n \left\| H_*^{-1}\tilde{L}_{g*}(-sI - \tilde{A}_{g*})^{-1}\tilde{B}_g \right\|_\infty.
\end{aligned}
$$

Because $\tilde{\Sigma} = \sigma_n I$, (2.3.37) and (2.3.41) yield that $\tilde{B}_g\tilde{B}_{g*} = \tilde{K}_{g*}\tilde{K}_g$ and so

$$
\begin{aligned}
\left\| G^{-1}(G - \hat{G}) \right\|_\infty &\le 2\sigma_n + \sigma_n \left\| H_*^{-1}\tilde{L}_{g*}(-sI - \tilde{A}_{g*})^{-1}\tilde{K}_{g*} \right\|_\infty \\
&= 2\sigma_n + \sigma_n \left\| H_*^{-1}(H_* - \hat{H}_*) \right\|_\infty \\
&= 2\sigma_n + \sigma_n \left\| (H - \hat{H})H^{-1} \right\|_\infty.
\end{aligned}
$$

Similarly,

$$\left\|\left(H - \hat{H}\right)^{-1} H^{-1}\right\|_\infty \le 2\sigma_n + \sigma_n \left\|G^{-1}\left(G - \hat{G}\right)\right\|_\infty.$$

The last two inequalities then yield (2.3.49). □

Our final task is to prove the error bound (2.2.9), applicable when the reduction is not of the specialized type considered above. With slight change of notation, let $\hat{G}_{N-1}, \hat{G}_{N-2}, \dots, \hat{G}_r$ denote the reduced order transfer function matrices obtained by eliminating all singular values equal to ν_N, or ν_N and $\nu_{N-1}, \dots,$ or ν_N and ν_{N-1} and $\dots, \nu_{r+1}$. Inequality (2.3.49) states that

$$\hat{G}_{N-1} = G\left(I - \frac{2\nu_N}{1-\nu_N}\delta_N\right),$$

where δ_N is a transfer function matrix (stable, as we have proved) with $\|\delta_N\|_\infty \le 1$. More generally, it is evident that

$$\hat{G}_r = G\left(I - \frac{2\nu_N}{1-\nu_N}\delta_N\right)\left(I - \frac{2\nu_{N-1}}{1-\nu_{N-1}}\delta_{N-1}\right)\cdots\left(I - \frac{2\nu_{r+1}}{1-\nu_{r+1}}\delta_{r+1}\right).$$

Evidently,

$$\begin{aligned}\left\|G^{-1}\left(G - \hat{G}_r\right)\right\|_\infty &\le \left(1 + \frac{2\nu_N}{1-\nu_N}\right)\cdots\left(1 + \frac{2\nu_{r+1}}{1-\nu_{r+1}}\right) - 1 \\ &= \prod_{r+1}^{N} \frac{1+\nu_i}{1-\nu_i} - 1.\end{aligned}$$

In Wang and Safonov (1992) the bound we have just established is extended to be a multiplicative error bound:

Lemma 2.3.8. *With notation as earlier, let $\hat{G}_r$ denote the reduced order transfer function obtained by eliminating all singular values of $F_c(s)$ equal to $\nu_N, \nu_{N-1}, \dots, \nu_{r+1}$, where $\nu_N < \nu_{N-1} < \cdots < \nu_{r+1} < \cdots$. Then*

$$\left\|\hat{G}_r^{-1}\left(G - \hat{G}_r\right)\right\|_\infty \le \prod_{r+1}^{N} \frac{1+\nu_i}{1-\nu_i} - 1. \tag{2.3.50}$$

Proof. Observe from (2.3.45) and then (2.3.46) that

$$\hat{H}_*^{-1}\tilde{C}_g = \tilde{K}_g + \hat{H}_*^{-1}L_{g*}\tilde{\Sigma} = \tilde{K}_g + \hat{H}_*^{-1}\hat{G}B_{g*}\tilde{\Sigma} + \hat{H}_*^{-1}\tilde{C}_g\tilde{\Sigma}^2$$

or

$$\hat{H}_*^{-1}\tilde{C}_g = \tilde{K}_g(I - \tilde{\Sigma}^2)^{-1} + \hat{H}_*^{-1}\hat{G}\tilde{B}_{g*}\tilde{\Sigma}(I - \tilde{\Sigma}^2)^{-1}.$$

Multiply by $\left(sI - \tilde{A}_g\right)^{-1}\tilde{B}_g$. There results

$$\begin{aligned}
&\hat{H}_*^{-1}(G - \hat{G}) \\
&\quad = \tilde{K}_g\left(I - \tilde{\Sigma}^2\right)^{-1}\left(sI - \tilde{A}_g\right)^{-1}\tilde{B}_g + \hat{H}_*^{-1}\hat{G}\tilde{B}_g^*\tilde{\Sigma}\left(I - \tilde{\Sigma}^2\right)^{-1}\left(sI - \tilde{A}_g\right)^{-1}\tilde{B}_g.
\end{aligned}$$

Now suppose that $r = N - 1$. Then

$$\begin{aligned}
&\left\|\hat{H}_*^{-1}(G - \hat{G}_{N-1})\right\| \\
&\qquad = \left\|\hat{G}_{N-1}^{-1}(G - \hat{G}_{N-1})\right\| \\
&\qquad \le \frac{1}{1-\nu_N^2}\left\|\tilde{K}_g\left(sI - \tilde{A}_g\right)^{-1}\tilde{B}_g\right\|_\infty + \frac{\nu_N}{1-\nu_N^2}\left\|\tilde{B}_g^*\left(sI - \tilde{A}_g\right)^{-1}\tilde{B}_g\right\|_\infty \\
&\qquad \le \frac{1}{1-\nu_N^2}2\nu_N + \frac{2\nu_N^2}{1-\nu_N^2} \quad \text{[using (2.3.37) and (2.3.41)]} \\
&\qquad = \frac{2\nu_N}{1-\nu_N}.
\end{aligned}$$

Suppose next to obtain an induction argument, that we have already established the inequality

$$\left\|\hat{G}_p^{-1}(G - \hat{G}_p)\right\|_\infty \le \prod_{p+1}^{N}\left(\frac{1+\nu_i}{1-\nu_i}\right) - 1. \tag{2.3.51}$$

Then

$$\hat{G}_{p-1}^{-1}(G - \hat{G}_{p-1}) = \hat{G}_{p-1}^{-1}\hat{G}_p[\hat{G}_p^{-1}(G - \hat{G}_p)] + \hat{G}_{p-1}^{-1}[\hat{G}_p - \hat{G}_{p-1}].$$

We have just demonstrated that

$$\left\|\hat{G}_{p-1}^{-1}(\hat{G}_p - \hat{G}_{p-1})\right\| \le \frac{2\nu_p}{1-\nu_p}$$

or that

$$\left\|\hat{G}_{p-1}^{-1}\hat{G}_p\right\| \le \frac{1+\nu_p}{1-\nu_p}.$$

Hence, appealing to the induction hypothesis (2.3.51)

$$\begin{aligned}
\left\|\hat{G}_{p-1}^{-1}(G - \hat{G}_{p-1})\right\|_\infty &\le \frac{1+\nu_p}{1-\nu_p}\left[\prod_{p+1}^{N}\left(\frac{1+\nu_i}{1-\nu_i}\right) - 1\right] + \frac{2\nu_p}{1-\nu_p} \\
&= \prod_{p}^{N}\left(\frac{1+\nu_i}{1-\nu_i}\right) - 1
\end{aligned}$$

and the induction step is established, as required. □

Variants on BST using singular perturbation

In our discussion in the previous chapter of balanced truncation, we exhibited a class of reduced order models obeying the same error bound formula as that for balanced truncation approximation. The reduced order models were obtained from a balanced realization by singular perturbation, or a generalization of it, and allowed the approximation error to be zero at any point, including 0, on the nonnegative real axis. [Balanced truncation corresponds to zero error at the point infinity].

It is also possible to vary BST in this way. We refer the reader to Green and Anderson (1990).

The algorithm proceeds as in Section 2.2 to the point where

$$\begin{bmatrix} Z(s) & G(s) \\ H(s) & F_c(s) \end{bmatrix} = \begin{bmatrix} D_z & D \\ D_h & 0 \end{bmatrix} + \begin{bmatrix} C \\ K \end{bmatrix} (sI - A)^{-1} \begin{bmatrix} L & B \end{bmatrix} \tag{2.3.52}$$

with

$$A = \begin{bmatrix} A_{11} & A_{12} \\ A_{21} & A_{22} \end{bmatrix}, \qquad B = \begin{bmatrix} B_1 \\ B_2 \end{bmatrix}, \qquad C = \begin{bmatrix} C_1 & C_2 \end{bmatrix} \quad \textit{etc.} \tag{2.3.53}$$

and the realization $\{A, B, K\}$ balanced. Instead of taking

$$\hat{G}_k(s) = D + C_1 (sI - A_{11})^{-1} B_1,$$

one takes, for any $\alpha \in [0, \infty]$,

$$\begin{aligned} \hat{A} &= A_{11} + A_{12}(\alpha I - A_{22})^{-1} A_{21}, & \hat{B} &= B_1 + A_{12}(\alpha I - A_{22})^{-1} B_2, \\ \hat{C} &= C_1 + C_2(\alpha I - A_{22})^{-1} A_{21}, & \hat{D} &= D + C_2(\alpha I - A_{22})^{-1} B_2. \end{aligned} \tag{2.3.54}$$

It results that $\hat{G}_k(\alpha) = G(\alpha)$, and the same error bound formula applies as earlier. Also, $\hat{G}_k(s)$ is of course stable.

Discrete-time BST

Discrete time BST works essentially the same way as continuous time BST. It is treated in Wang and Safonov (1991). Of course, the form of the Lyapunov and Riccati equations is somewhat changed. In Green and Anderson (1990) it is pointed out that an equivalent to discrete time balanced stochastic truncation of $G_d(z)$ to yield $\hat{G}_{dk}(z)$ is the following. From $G_d(z)$ form $G(s) = G_d((1+s)/(1-s))$. Construct $\hat{G}_k(s)$ by generalized BST with $G(1) = \hat{G}_k(1)$, and set $\hat{G}_{dk}(z) = \hat{G}_k((z-1)/(z+1))$. The same error bound applies—it is established directly in Wang and Safonov (1991) and using the generalized BST result in Green and Anderson (1990).

Main points of the section

1. The claims made in the overview of the BST algorithm are all established. This includes claims connecting certain transfer functions and state-variable realizations, and lower and upper error bounds.

2. The number of right half plane zeros of the transfer function $G(s)$ to be approximated equals the number of Hankel singular values of $F_c(s)$ which are equal to 1. Here, $F_c(s)$ is the transfer function subject to (additive) balanced truncation within the algorithm.

3. For a minimum phase $G(s)$ the algorithm can be simplified, replacing a Riccati equation by a Lyapunov equation.

4. The algorithm can be extended to nonsquare transfer function matrices, and also to give a generalized balanced stochastic truncation, ensuring zero error at an arbitrary point on the extended nonnegative real axis.

5. Discrete time balanced stochastic truncation is possible, and is trivially different to continuous time BST.

2.4 Multiplicative Hankel Norm Approximation

Balanced stochastic truncation uses within it balanced truncation (in the sense of the last chapter). Analogously, multiplicative Hankel norm approximation uses (additive) Hankel norm approximation within it. Indeed, BST and multiplicative Hankel norm reductions are executed in almost the same way, the only significant difference being in the (additive error) reduction of the transfer function matrix we have termed $F_c(s)$, the causal part of a certain all-pass.

The method is originally to be found in Glover (1986). For interesting connections with balanced stochastic truncation, see Green (1988b). For some further developments and properties, see Matson, Lam, Anderson and James (1993).

In additive error Hankel norm reduction, a choice is available of reducing a $G(s)$ to $\hat{G}_k(s)$ in one step, or in a series of steps, with a sequence of intermediate approximates of decreasing degrees; if all Hankel singular values of $G(s)$ are distinct, the approximants are in fact $\hat{G}_{n-1}, \hat{G}_{n-2}, \ldots, \hat{G}_k$.

In multiplicative Hankel norm approximation, we again encounter a series of steps, each step corresponding to the elimination of a distinct, but possibly repeated, Hankel singular value.

There is one aspect in which the multiplicative Hankel norm problem is potentially more general than the BST problem: the object we are trying to reduce may be unstable. Because of this, it is helpful to review what exactly we are trying to do in the approximation process.

Let $S(n_1, n_2)$ denote the class of rational transfer function matrices with n_1 poles in $\mathrm{Re}[s] < 0$ and n_2 poles in $\mathrm{Re}[s] \geq 0$. Then the natural problem to consider is: for given $G \in S(n_1, n_2)$ find $\hat{G}_k \in S(k, n_2)$ such that

$$\hat{G}_k = G(I - \Delta), \tag{2.4.1}$$

with $\|\Delta\|$ as small as possible. We are not requiring Δ to be stable. It will turn out that when an analytic solution to the minimization problem is available, Δ can be taken as stable with no loss of generality. Also, in suboptimal solutions, we shall take Δ to be stable.

A lower bound on the approximation error

Suppose that G has full row rank. Then there exists a square $H(s)$ such that $H(s)$ and $H^{-1}(s)$ have no poles in $\mathrm{Re}[s] > 0$ and

$$GG_* = H_* H. \tag{2.4.2}$$

Note that H and H^{-1} may have pure imaginary poles. Also $(H_*^{-1}G)(H_*^{-1}G)_* = I$. We claim that choosing Δ such that in (2.4.1), $\hat{G}_k \in S(k, n_2)$ and $\|\Delta\|$ is minimized is equivalent to choosing $\hat{G}_k \in S(k, n_2)$ such that $\|H_*^{-1}(G - \hat{G}_k)\|_\infty$ is minimized. For if $\hat{G}_k$ yields minimum $\|\Delta\|_\infty$ for (2.4.1), then

$$H_*^{-1}\left(G - \hat{G}_k\right) = H_*^{-1} G \Delta \quad \text{and} \quad \left\| H_*^{-1}\left(G - \hat{G}_k\right) \right\|_\infty \leq \|\Delta\|_\infty .$$

[With $H_*^{-1}G$ not necessarily square, the inequality necessarily results]. Conversely, if $\hat{G}_k \in S(k, n_2)$ yields minimum $\|H_*^{-1}(G - \hat{G}_k)\|_\infty$ define $E = H_*^{-1}(G - \hat{G}_k)$ and $\Delta = G_* H^{-1} E$. Then $\|\Delta\|_\infty \leq \|E\|_\infty$ and $G(I - \Delta) = \hat{G}_k$.

Hence the two minimization problems are equivalent. Now observe that

$$\inf_{\hat{G}_k \in S(k, n_2)} \left\| H_*^{-1}\left(G - \hat{G}_k\right) \right\| \geq \inf_{X \in S(k, \infty)} \left\| H_*^{-1}\left(G - X\right) \right\|_\infty = \sigma_{k+1}\left(\left[H_*^{-1}G\right]_+\right)$$

or

$$\|\Delta\| \geq \sigma_{k+1}(F_c). \tag{2.4.3}$$

Here, F_c is, as before, the causal part of $F = H_*^{-1}G$ in an additive decomposition of F.

Overview of reduction approach—eliminating one Hankel singular value only

Suppose $F_c(s)$ has Hankel singular values $1 \geq \sigma_1 \geq \sigma_2 \geq \cdots \geq \sigma_{n_1}$ where n_1 is the number of poles of G in the open left half plane. Suppose that σ_{n_1} has multiplicity l.

Then, as we know, there exists $\hat{F}_c(s)$, a stable transfer function matrix of degree $n_1 - l$, such that

$$F_c(s) - \hat{F}_c(s) = E(s) \tag{2.4.4}$$

obeys

$$E(s)E_*(s) = \sigma_{n_1}^2 I. \tag{2.4.5}$$

Then it turns out that an approximation $\hat{G}(s)$ of $G(s)$ with $n_1 - l$ strictly stable poles and the same number of closed right half plane poles as G results from setting

$$\hat{G} = G - H_* E. \tag{2.4.6}$$

(In case $1 = \sigma_1 = \cdots = \sigma_{n_1}$, no useful approximation can be performed).

It is not *a priori* obvious that $\hat{G}$ has the claimed pole distribution property. Assume this to be the case for the moment; we will establish the claim later. Observe that

$$\hat{G} = G - H_* E = G - GG_*\left(H_* H\right)^{-1} H_* E = G\left(I - G_* H^{-1} E\right). \tag{2.4.7}$$

Setting

$$\Delta = G_* H^{-1} E, \tag{2.4.8}$$

we see that

$$\Delta_* \Delta = E_* H_*^{-1} G G_* H^{-1} E = E_* E. \tag{2.4.9}$$

Hence

$$\|\Delta\|_\infty = \sigma_{n_1}. \tag{2.4.10}$$

In the square $G(s)$ case, the transfer function matrix $E(s)$ has several further useful properties. First, it ensures that the transfer function matrix

$$\hat{H} = H - EG_* \tag{2.4.11}$$

turns out to be stable and minimum phase (apart from allowing poles and zeros on the imaginary axis), and satisfying

$$\hat{H}_* \hat{H} = \hat{G}\hat{G}_*. \tag{2.4.12}$$

A second property of E in the square case demonstrated later is that

$$\left[\hat{H}_*^{-1}\hat{G}\right]_+ = \hat{F}_c. \tag{2.4.13}$$

Now recall from the earlier discussion on additive Hankel norm approximation that the Hankel singular values of $\hat{F}_c$ are the same as those of F_c, except for the l repeats of the smallest Hankel singular value. This is the same thing as saying that if $\nu_1(G), \ldots, \nu_N(G)$ are the distinct Hankel singular values of $F_c(s)$ (of course they are ultimately determined by G), then

$$\nu_i(\hat{G}) = \nu_i(G), \qquad i = 1, \ldots, N-1. \tag{2.4.14}$$

As far as the actual algorithm is concerned, in broad outline, it mimics that for BST up to the point of finding $F_c(s)$. Of course, we must allow for $G(s)$ to have poles or zeros in $\text{Re}[s] \geq 0$. Then $F_c(s)$ is reduced in a new way, by Hankel norm approximation. Finally, $\hat{G}$ is obtained in a new way. Later, we indicate a state-variable formula for $\hat{G}$.

Overview of reduction approach—general case and the error bound formula

Above, we have described one step of what may be a multistep process when a $\hat{G}_k$ of sufficiently small k is needed.

Now we have remarked that if $G(s)$ is square, when the same calculations are repeated on $\hat{G}$ as we have performed on G, and we form an all-pass to replace $F(s)$ and the causal part of this all-pass to replace $F_c(s)$, this last transfer function matrix has Hankel singular values $\sigma_1, \sigma_2, \ldots, \sigma_{n_1-l}$, *i.e.*, it has the same Hankel singular values as resulted from starting with G, except for the last l. If $G(s)$ is not square, it turns out that the original Hankel singular values down to σ_{n_1-l} overbound those derived from $\hat{G}(s)$. This means that if there are repeated Hankel singular values in the list $\sigma_1, \sigma_2, \ldots, \sigma_{n-l}$ they will *not* in general be repeated in the list $\hat{\sigma}_1, \hat{\sigma}_2, \ldots, \hat{\sigma}_{n-l}$ associated with $\hat{G}(s)$. Now one step in the reduction process removes (all occurrences of) the smallest Hankel singular value in the list. Repeated singular values are therefore an advantage, in terms of the number of steps required, but this advantage is likely to be lost after the first step for a nonsquare $G(s)$.

There is an alternative. This works by augmenting $G(s)$ through the addition of extra rows to make it square, but in the process not increasing the degree or the relevant Hankel singular values, see Glover (1986) for details, as well as the previous section. Then one can reduce the square augmented $G(s)$. At the end of the procedure, the extra rows are deleted to give the nonsquare approximation to the original nonsquare $G(s)$.

This procedure, although apparently clumsy, does not destroy any multiplicities of Hankel singular values, and so may involve fewer steps than the first possibility for nonsquare $G(s)$.

The alternative exploiting multiple singular values has the potential to yield a smaller error bound, as we shall indicate below.

Disregarding for the moment how we may be dealing with nonsquare $G(s)$, observe that after one reduction step, a further reduction step is possible if the new smallest Hankel singular value of the new $F_c(s)$ is smaller than one. If $\nu_1, \nu_2, \ldots, \nu_N$ denote in descending order the distinct Hankel singular values of $F_c(s)$, and if after the $N - rth$ reduction step the replacement of $F_c(s)$ has distinct Hankel singular values $\nu_1, \nu_2, \ldots, \nu_r$ with $\sigma_k = \nu_r$ and $\sigma_{k+1} = \nu_{r+1}$, we have an approximant

$$\hat{G}_k = G\,(I - \nu_N\delta_N)\ldots(I - \nu_{r+1}\delta_{r+1}), \tag{2.4.15}$$

where the δ_i are stable and with maximum singular value bounded by 1 on the imaginary axis. (The stability of the δ_i will be argued later.) This equation is valid both when G is square, or when it is replaced by an augmented square quantity. When it is not square the ν_i for $i < N$ should be replaced by new values which are smaller than or equal to the ν_i associated with the original $G(s)$.

Evidently, there holds

$$\left\| G^{-1}\left(G - \hat{G}_k\right)\right\|_\infty \leq (1 + \nu_{r+1})\ldots(1 + \nu_N) - 1. \tag{2.4.16}$$

Further, since the δ_i are stable with norm bound of 1 and the ν_i appearing in the formula (2.4.15) are all less than one, each product $I - \nu_i\delta_i$ has all its poles and the poles of its inverse in the open left half plane. As a consequence, $\hat{G}_k(s)$ inherits the closed right half plane poles and zeros of $G(s)$. This observation implies that there is a least degree for $\hat{G}_k(s)$, which may be greater than 1.

The error bounds for the two approaches to nonsquare $G(s)$ reduction will differ in that, in the first method, the σ_i will, apart from the last, generally be smaller than those for the second method, based on augmenting $G(s)$ to make it square. However, if the original $G(s)$ has repeated σ_i, there will in the second method be fewer terms of the form $(1 + \nu_i)$ which are multiplied together to produce the error bound. Consequently, the error bound may be smaller.

The error bound (2.4.17) should be compared with that applying for balanced stochastic truncation. In fact, each multiplicand $(1 + \nu_i)$ on the right in (2.4.17) has to be replaced by a multiplicand $(1 + \nu_i)(1 - \nu_i)^{-1}$ in the balanced stochastic truncation error bound formula. Evidently then, the multiplicative Hankel norm reduction error bound is smaller than the BST error bound.

In the next subsection, we give more details on the actual algorithm required to execute the reduction. Of course, it is a sequential one, removing so to speak the smallest singular value (in its full multiplicity) at each reduction step.

Algorithm for reduction

We shall assume from the beginning that $G(s)$ satisfies

$$G(\infty)G^T(\infty) = I. \tag{2.4.17}$$

If G is singular at $s = \infty$, or has a pole at $s = \infty$, preliminary transformations as set out in Glover (1986) will replace the original problem by one satisfying $G(\infty)G^T(\infty) = I$. The most comprehensive algorithm description is to be found in Matson *et al.* (1993).

Step 1. Construction of the stable minimum phase spectral factor $H(s)$. This construction results in a square $H(s)$ which has no poles or zeros in the open right half plane; it may have poles or zeros on the imaginary axis. Suppose $G(s)$ has a minimal realization

$$G(s) = \left[\begin{array}{cc|c} A_1 & 0 & B_1 \\ A_{21} & A_2 & B_2 \\ \hline C_1 & C_2 & D \end{array}\right], \tag{2.4.18}$$

with $DD^T = I$, all eigenvalues of A_1 in the open left half plane and all eigenvalues of A_2 in the closed right half plane. Then $H(s)$ is

$$H(s) = \left[\begin{array}{c|c} A & L \\ \hline K & I \end{array}\right]. \tag{2.4.19}$$

The constituent matrices of $H(s)$ are found in the following way. Let P be the controllability grammian of the stable part of $G(s)$:

$$A_1 P + P A_1^T + B_1 B_1^T = 0. \tag{2.4.20}$$

Set

$$A = \begin{bmatrix} A_1 & -(PA_{21}^T + B_1 B_2^T) \\ 0 & -A_2^T \end{bmatrix}, \qquad L = \begin{bmatrix} PC_1^T + B_1 D^T \\ C_2^T \end{bmatrix}. \tag{2.4.21}$$

Next, let

$$\tilde{B} = \begin{bmatrix} P^{-1} B_1 \\ B_2 \end{bmatrix} \tag{2.4.22}$$

and let X be the strong solution of the Riccati equation

$$(A + LD\tilde{B}^T)^T X + X(A + LD\tilde{B}^T) - XLL^T X + \tilde{B} D_\perp^T D_\perp \tilde{B}^T = 0 \tag{2.4.23}$$

where $D_\perp$ is such that $\left[\begin{smallmatrix} D \\ D_\perp \end{smallmatrix}\right]$ is unitary. (The strong solution is one for which $A + LD\tilde{B}^T - LL^T X$ has eigenvalues in the closed left half plane rather than the open left half plane.) Then

$$K = L^T X - D\tilde{B}^T. \tag{2.4.24}$$

The fact that $H_* H = GG_*$ and H has the required analyticity properties is established in Matson *et al.* (1993)

The proof is straightforward, using of course (2.4.20) and (2.4.23) in evaluating $H_* H$ and $G_* G$. The analyticity properties of H and H^{-1} are a consequence of (2.4.21) and the fact that X is a strong solution to (2.4.23). The zeros of H are the eigenvalues of $A + LD\tilde{B}^T - LL^T X$.

Step 2. Construction of the stable part of $H_*^{-1}G$. As shown in Matson *et al.* (1993), there holds

$$F_c(s) = \left[H_*^{-1}G\right]_+ = \begin{bmatrix} A_1 & B_1 \\ K_1 & D \end{bmatrix}. \tag{2.4.25}$$

Of course, $K = [K_1 \; K_2]$, partitioned conformably with A. Equation (2.4.25) is most easily derived by evaluating $H_* F_c$ and checking that its stable part is identical with the stable part of G.

Step 3. Reduction of $F_c(s)$ to secure an all-pass error. Suppose that a transformation matrix T is found so that $A_1^b = T^{-1}A_1T$, $B_1^b = T^{-1}B_1$ and $K_1^b = KT$ are in a balanced form with observability and controllability grammians given by

$$\Sigma = \operatorname{diag}(\hat{\Sigma},\ \sigma_{n_1} I_l) \tag{2.4.26}$$

and

$$A_1^b = \begin{bmatrix} \bar{A}_{11} & \bar{A}_{12} \\ \bar{A}_{21} & \bar{A}_{22} \end{bmatrix}, \qquad B_1^b = \begin{bmatrix} \bar{B}_1 \\ \bar{B}_2 \end{bmatrix}, \qquad K_1^b = \begin{bmatrix} \bar{K}_1 & \bar{K}_2 \end{bmatrix}. \tag{2.4.27}$$

Of course, σ_{n_1} is the smallest Hankel singular value, with multiplicity l. Set

$$\begin{aligned}
\hat{A}_1 &= \left(\hat{\Sigma}^2 - \sigma_{n_1}^2 I\right)^{-1}\left(\sigma_{n_1}^2 \bar{A}_{11}^T + \hat{\Sigma}\bar{A}_{11}\hat{\Sigma} - \sigma_{n_1}\bar{K}_1^T U \bar{B}_1^T\right), \\
\hat{B}_1 &= \left(\hat{\Sigma}^2 - \sigma_{n_1}^2 I\right)^{-1}\left(\hat{\Sigma}\bar{B}_1 + \sigma_{n_1}\bar{K}_1^T U\right), \\
\hat{K}_1 &= \left(\bar{K}_1\hat{\Sigma} + \sigma_{n_1} U \bar{B}_1^T\right), \\
\hat{D} &= D - \sigma_{n_1} U,
\end{aligned} \tag{2.4.28}$$

where U is a matrix such that $UU^T = I$ and $\bar{B}_2 + \bar{K}_2^T U = 0$. (Such a matrix always exists.)

Then

$$\hat{F}_c(s) = \begin{bmatrix} \hat{A}_1 & \hat{B}_1 \\ \hat{K}_1 & \hat{D} \end{bmatrix}. \tag{2.4.29}$$

(This ensures that (2.4.4) and (2.4.5) hold.)

Step 4. Construction of $\hat{G}(s)$ for a single step approximation Introduce first the following transformed state variable realizations of $G(s)$ and $H(s)$:

$$G(s) = \left[\begin{array}{cc|c} T^{-1}A_1T & 0 & T^{-1}B_1 \\ A_{21}T & A_2 & B_2 \\ \hline C_1T & C_2 & D \end{array}\right] = \left[\begin{array}{cc|c} A_1^b & 0 & B_1^b \\ A_{21}^b & A_2 & B_2 \\ \hline C_1^b & C_2 & D \end{array}\right], \tag{2.4.30}$$

$$H(s) = \left[\begin{array}{cc|c} T^{-1}A_1T & T^{-1}A_{12} & T^{-1}L_1 \\ 0 & -A_2^T & C_2^T \\ \hline K_1T & K_2 & I \end{array}\right] = \left[\begin{array}{cc|c} A_1^b & A_{12}^b & L_1^b \\ 0 & -A_2^T & C_2^T \\ \hline K_{1b} & K_2 & I \end{array}\right], \tag{2.4.31}$$

with $A_{12} = -(PA_{21}^T + B_1B_2^T)$ as in (2.4.21). Finally,

$$\hat{G}(s) = \left[\begin{array}{cc|c} \hat{A}_1 & 0 & \hat{B}_1 \\ \hat{A}_{21} & A_2 & \hat{B}_2 \\ \hline \hat{C}_1 & C_2 & \hat{D} \end{array}\right]. \tag{2.4.32}$$

Here A_2 and C_2 are unchanged from the original system, and the other quantities are defined as

$$\begin{aligned} \hat{A}_{21} &= -(A_{12}^{bT}\begin{bmatrix} \hat{\Sigma}^2 - \sigma_{n_1}^2 I \\ 0 \end{bmatrix} + K_2^T\hat{K}_1), \\ \hat{C}_1 &= \hat{K}_1 + L_1^{bT}\begin{bmatrix} \hat{\Sigma}^2 - \sigma_{n_1}^2 I \\ 0 \end{bmatrix}, \\ \hat{B}_2 &= B_2 + \sigma_{n_1}K_2^TU - X_{12}^TB_1, \\ \hat{D} &= D - \sigma_{n_1}U, \end{aligned} \tag{2.4.33}$$

where U is as earlier and X_{12} is a submatrix of X in (2.4.23):

$$X = \begin{bmatrix} X_{11} & X_{12} \\ X_{12}^T & X_{22} \end{bmatrix}. \tag{2.4.34}$$

In Matson *et al.* (1993), a numerically stable algorithm is presented, which avoids the numerically undesirable step of constructing a balanced realization.

In the earlier part of the section, we made a number of claims about various quantities arising in the calculations. Following are remarks concerning the proof of those claims whose proof has not yet been discussed.

In Matson *et al.* (1993) it is also proved by straightforward but lengthy algebraic manipulation that the $\hat{G}$ of (2.4.30) does in fact satisfy

$$\hat{G} = G - H_*E.$$

The unstable eigenvalues of $\hat{A}$ are the same as the unstable eigenvalues of A, *viz.* eigenvalues of A_2. Thus $\hat{G}$ has the requisite pole properties—in fact it inherits the unstable poles of G. It also inherits the unstable zeros, as in the BST case.

Similarly, lengthy algebra will establish that $F_* E = G_* H^{-1} E$ is stable.

In the square case, it is quick to verify that the definition of $\hat{H}$ in (2.4.11) ensures that $\hat{H}_* \hat{H} = \hat{G}\hat{G}_*$—one just needs $EE_* = E_* E = \sigma_{n_1}^2 I$ and $GG_* = H_* H$. Not only is $F_* E$ stable but so also is EF_*. Then by (2.4.11)

$$\hat{H} = H - EG_* H^{-1} H = (I - EF_*)\, H.$$

So $\hat{H}$ is stable. Because $\|EF_*\|_\infty = \sigma_{n_1} < 1$, $(I - EF_*)^{-1}$ is stable and so $\hat{H}^{-1}$ is stable.

In the square case, one can verify by a direct calculation that $GE_* \hat{F}_c$ is unstable. Then observe that

$$\begin{aligned}
\hat{H}_* \hat{F}_c &= (H_* - GE_*)\hat{F}_c && \text{[using (2.4.11)]} \\
&= H_* F_c - H_* E - GE_* \hat{F}_c && \text{[using (2.4.4)]} \\
&= \left[H_* F - H_*(F - F_c)\right] + \left[\hat{G} - G\right] - \text{unstable} \\
& && \text{[using (2.4.6) and the instability of } GE_* \hat{F}_c] \\
&= \left[G - \text{unstable}\right] + \left[\hat{G} - G\right] - \text{unstable}, \\
& && \text{[using the fact that } F_c \text{ is the causal part of } F = H_*^{-1} G]
\end{aligned}$$

i.e.,

$$\hat{F}_c = \hat{H}_*^{-1}\hat{G} + \text{unstable} \quad \text{or} \quad \hat{F}_c = \left[\hat{H}_*^{-1}\hat{G}\right]_+.$$

This is the basis for the claim that $\nu_i(\hat{G}) = \nu_i(G)$, $i = 1, \ldots, N-1$ in the square case. In the nonsquare case, $\nu_i(\hat{G}) \leq \nu_i(G)$. This argument is given in Matson *et al.* (1993).

The discussion in terms of $\hat{H}$ has not really drawn on its state-variable description. In Glover (1986), proofs are given of virtually all the claims above without direct appeal to state-space descriptions, although some key unproven lemmas will depend for their proof on such descriptions.

Main points of the section

1. For unstable and nonsquare but full row rank G, one can perform multiplicative Hankel norm reduction.

2. The key is to use Hankel norm reduction on the stable part of $H_*^{-1} G$ where $H_* H = GG_*$, with H and H^{-1} analytic in $\text{Re}[s] > 0$.

3. Reduction proceeds by eliminating one distinct Hankel singular value of $[H_*^{-1} G]_+$, at a time, and each step individually is optimal.

4. The error bound formula is $\prod_{k+1}^{N}(1+\nu_i)-1$ and is tighter than the BST error bound formula.

5. The reduced order $\hat{G}$ has the same unstable poles and zeros as G.

2.5 Example

An example is given here to illustrate the difference between additive error and multiplicative error criteria. Consider the following seventh order transfer function, Safonov and Chiang (1988):

$$G(s) = \frac{0.05(s^7+801s^6+1024s^5+599s^4+451s^3+119s^2+49s+5.55)}{s^\tau+12.6s^6+53.48s^5+90.94s^4+71.83s^3+27.22s^2+4.75s+0.3}$$

Let us obtain four third order models: $G_{bst}(s)$, $G_{mk}(s)$, $G_{bt}(s)$, $G_{oh}(s)$ using balanced stochastic truncation, multiplicative Hankel norm approximation, balanced truncation, and optimal Hankel norm approximation, respectively. $G_{bst}(s)$ and $G_{mh}(s)$ are based on a multiplicative error criterion while $G_{bt}(s)$ and $G_{oh}(s)$ are based on an additive error criterion. The frequency responses are compared in Figure 2.5.1 and Figure 2.5.2. $G_{bst}(s)$ and $G_{mh}(s)$ are evidently better than the others in a wider frequency range. Especially, the goodness of fit is clear in the phase characteristics.

Figure 2.5.3 shows the multiplicative errors with the H_∞ norm bounds for $G_{bst}(s)$ and $G_{mh}(s)$, given by (2.2.9) and (2.4.16).

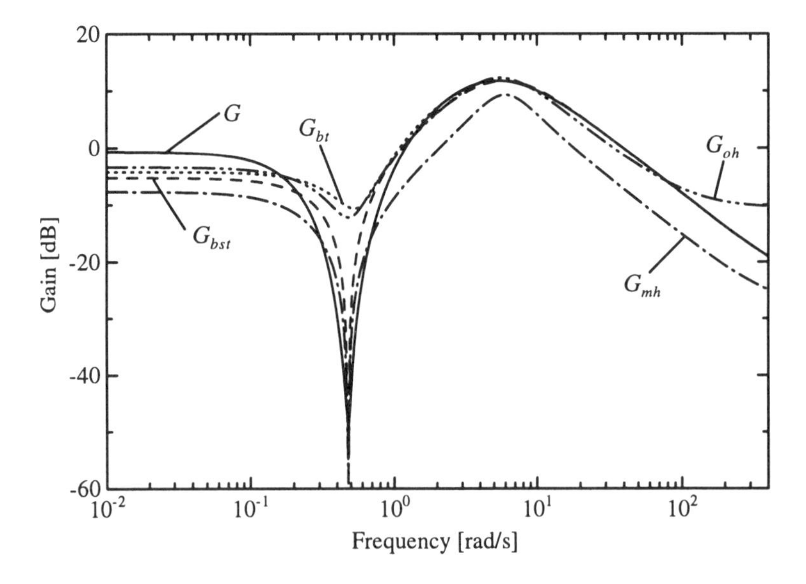

Figure 2.5.1. Comparison of gain characteristics for two multiplicative and two additive error reductions

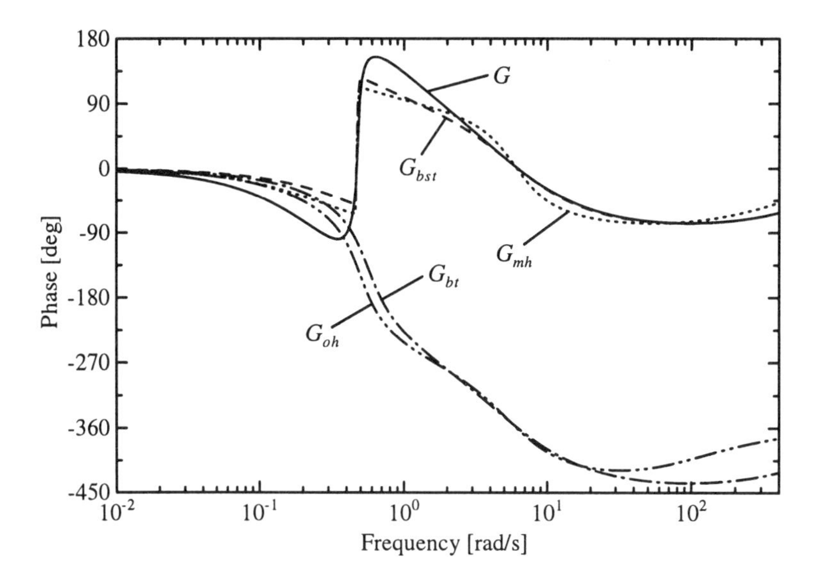

Figure 2.5.2. Comparison of phase characteristics for two multiplicative and two additive error reductions

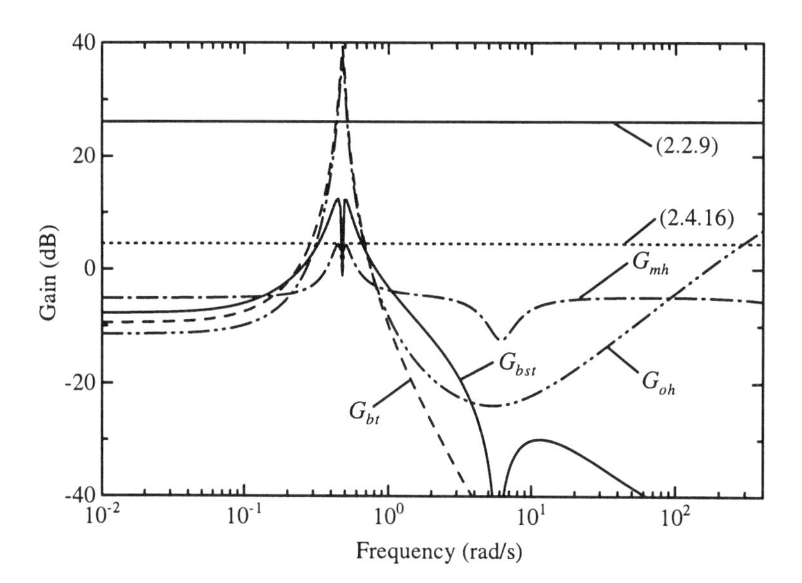

Figure 2.5.3. Multiplicative error and H_∞ error bounds for the four reduction methods

Chapter 3

Low Order Controller Design

3.1 Approaches to the Design of Low Order Controllers

Simple linear controllers are normally to be preferred to complex linear controllers for linear time-invariant plants. There are fewer things to go wrong in the hardware or bugs to fix in the software, they are easier to understand, and the computational requirements (and associated hardware requirements) are less. On the other hand, many plants are of high order, Therefore, there is a desire to have methods available for the design of controllers which have order significantly less than the order of the plant they are controlling.

Unfortunately, the two most widely used analytical controller design procedures, namely H_2 or linear-quadratic-gaussian (LQG) and H_∞, normally yield controllers with order equal to or approximately equal to the order of the plant. These methods, although analytical in character, have now reached a stage of refinement that it is now possible to incorporate many classical and semi-quantitative objectives (*e.g.*, maintain low coupling between channels) in the design process, and so are deservedly popular.

How then may low order controller designs be achieved for high order plants? The methods can broadly be divided into classes: direct, in which the parameters defining a low-order controller are computed by some optimization or other procedure, and two types of indirect design; in one of these a high-order controller is first found and then a procedure used to simplify it, and in the other the high order plant model is approximated by a low order model, and then a low order controller is found using this approximating low order plant model, but of course used in a loop with the real high order plant. Figure 3.1.1 illustrates the concepts.

Direct design methods

Examples of direct methods include the work of Gangsaas, Bruce, Blight and Ly (1986), see especially the third case study in this paper, and Bernstein and Hyland

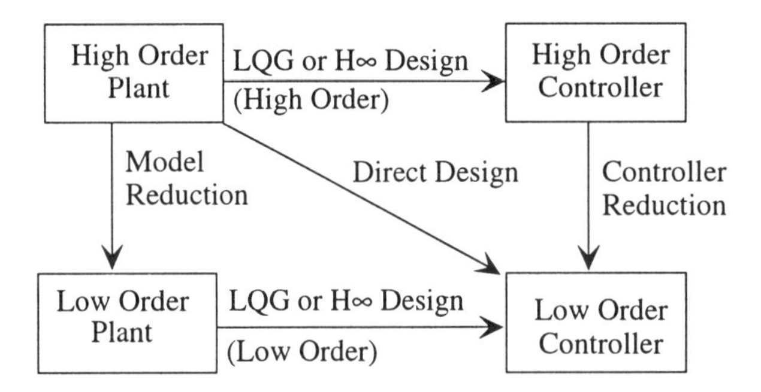

Figure 3.1.1. Approaches to low order controller design for a high order plant

(1985). This book does not have such methods as a central concern, but they nevertheless demand certain comment. The common philosophy of such methods is to seek to minimize a performance index, often a quadratic performance index, subject to the constraint that the controller be of fixed degree (as well as being stabilizing and time-invariant).

The central idea of the scheme in Gangsaas *et al.* (1986) is to compute the sensitivity of the performance index to a variation of a control parameter, and to work with a performance index computed over a long but finite time-interval, which allows the closed-loop stability problem to be indirectly addressed; of course the controller parameters are iterated till the index is minimized. There can be no assurance that a global minimum is reached.

The strategy used in Bernstein and Hyland (1985) and many related works is to determine a set of algebraic equations which constitute necessary but not sufficient conditions on the controller parameters to achieve a minimum value for the performance index. These equations have a structure which displays a parallel with those applicable in the full-order case, and the number of (scalar) equations equals the number of scalar controller parameters. The solution of these equations is far from straightforward, and the multiplicity of solutions may be very great. Nevertheless, progress is being made using homotopic methods, see *e.g.*, Greeley and Hyland (1987). As far as is known, there are no commercially available software packages offering this design procedure. On the other hand there are commercially available software packages that will allow low order controller design through reduction of controllers of order as high as 55; a problem of this dimension would represent a very sizeable challenge to any direct design method based on solving simultaneous algebraic equations for the unknown controller parameters.

Another possible disadvantage of direct low-order design is the fact that although the conventional LQG design method is often used iteratively, through tuning of the weighting matrices, the extent to which this might be possible, and how to do it, in a context of direct low-order controller design is far from clear. Is it legitimate or not to mimic the tuning concept of loop transfer recovery, in pursuing a low-order design? No published answer is known.

Indirect methods

Most of this chapter deals with indirect methods. In this subsection, we simply want to summarize a few general issues.

First, it is misleading to think of controller reduction as something which can be achieved by picking out one of the model reduction procedures presented in the first chapter. Why? Consider how one might want to measure the quality of the approximation. Our goal should be to have the closed loop comprising the true plant with high order controller behave like the closed loop comprising the true plant with the low order controller. At the very least, we want to retain stability; but we might want to retain bandwidth, form of step response, value of a quadratic performance index, or we might want the two closed-loop transfer function matrices to be close to one another. Therefore we should have an approximation measure that ascribes high quality of approximation when a particular aspect of closed-loop behaviour is well matched. Whatever criterion we choose to reflect closeness of closed-loop behaviour however, *it is clear that the plant must necessarily be reflected somehow in the approximation criterion*. This is a fundamental difference with the situation considered in the earlier chapters.

Second, it should be fundamentally more attractive to design a high order controller first and approximate it, as opposed to first approximating a high order plant with a low order plant, and then designing a low order controller using the low order approximating plant, with the intention of connecting the controller to the original high order plant. Why? There is a general reason and a specific reason. The general reason is that if a design procedure involves approximation at some stage, it is normally better to have the approximation occur as late as possible; if it occurs early, it is possible that later stages in the design process can "massage" the approximation in an unpredictable way, so that what starts out as a low-error approximation can become a high-error one in the end. Approximating the plant and then designing the controller puts the approximation step earlier than if one designs a controller and then approximates that controller. The specific reason relates to the point made in the paragraph above. There we argued that logically-based controller approximation demands knowledge of the plant; the same argument (based on the desire to match closed-loop behaviour) indicates that plant approximation should best proceed with knowledge of the controller. But now one is well and truly impaled on the horns of a dilemma: one needs the reduced order plant to obtain the controller, and thus the closed loop, and one needs the controller to obtain the reduced order plant.

It turns out that there is a partial way out of the dilemma mentioned earlier, and we shall discuss that later in this chapter. We should note that the way out does not rely simply on using balanced truncation, or Hankel norm reduction, on a high order model of the plant, which was an early suggestion in the literature Glover and Limebeer (1983). Such an approach is likely to give a very poor design. Another way out of the dilemma, not discussed below, is to use an iterative approximation and design procedure. After each controller design, a low order approximation of the plant is recalculated using the new controller design and in a manner which seeks to match closed-loop behaviours; the new low order plant approximation is used to obtain yet another controller, and so on.

As mentioned above, it is in general difficult to obtain a good reduced-order controller via the indirect method of first approximating a high order plant and then defining a controller for this approximating plant. However there is one case where the construction of a reduced-order plant model by mode truncation, followed then by controller design, may lead to a good controller for the full order plant. Let

$$\left\{A_r = LAL^T\left(LL^T\right)^{-1},\ B_r = LB,\ C_r = CL^T\left(LL^T\right)^{-1}\right\}$$

denote a reduced order model of $\{A, B, C\}$ obtained by modal truncation. The state of the reduced order plant model $z(t)$ is related to the state of the original plant $x(t)$ by $z(t) = Lx(t)$. Any state feedback control $u = -Kz = -KLx$ has the effect of leaving unchanged those modes of the original plant thrown away in the truncation process (because $[A, L]$ is unobservable). Now an exact observer of z could be constructed from the input signal u and the signal $C_r z = CL^T(LL^T)^{-1}Lx$. If this observer is actually driven by Cx, as will happen where one attempts to attach the observer/state-feedback law based controller to the full-order system—unless additional sensors can be introduced, then one will no longer have exactly $u = -K\hat{z} = -KL\hat{x}$. Nevertheless, the error may be acceptable in terms of closed-loop performance (and in the special case that range L contains range C, the error is zero); the issue is whether models which are very small or negligible in open loop remain so in closed loop. The reduction procedure is described in Rao and Lamba (1975). A successful example applied to a vibration suppression problem appears in Seto and Mitsuta (1991).

Main points of the section

1. Low order controller design is described. Direct low order controller design for a high order plant is extremely difficult.

2. It is better to design a high order controller and then approximate it than to approximate a high order plant and design a low-order controller for the approximation.

3. Controller reduction must aim at preserving closed-loop behaviour, and so the plant is relevant to the controller reduction process.

3.2 Controller and Plant Reduction via Frequency Weighted Approximation

Most of this section, and indeed this chapter, will treat controller reduction rather than plant reduction, for the reasons set out in the last section.

We have argued in the previous section that controller reduction should seek to preserve properties of the closed loop in which the controller is located. In this section we aim to sharpen up this idea; the end result is to obtain some frequency weighted approximation problems. The original idea is due to Enns, see *e.g.*, Enns (1984). For more developments, see Anderson and Liu (1989)

Stability margin considerations for frequency weighting

Let $P(s)$ be the transfer function matrix of a linear time-invariant plant, and let $K(s)$ be a stabilizing high order controller (The procedure used to obtain $K(s)$ from $P(s)$ is not relevant at this stage.) Let $\hat{K}(s)$ be a reduced order controller, which we are seeking. Regard the closed-loop system with $\hat{K}(s)$ replacing $K(s)$ as being equivalent to Figure 3.2.1. Figures 3.2.2 and 3.2.3 show modifications of Figure 3.2.1 which have the same stability properties.

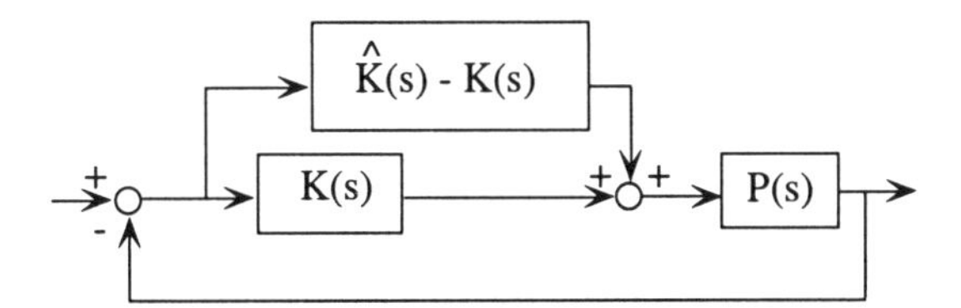

Figure 3.2.1. Effect of replacing K by $\hat{K}$

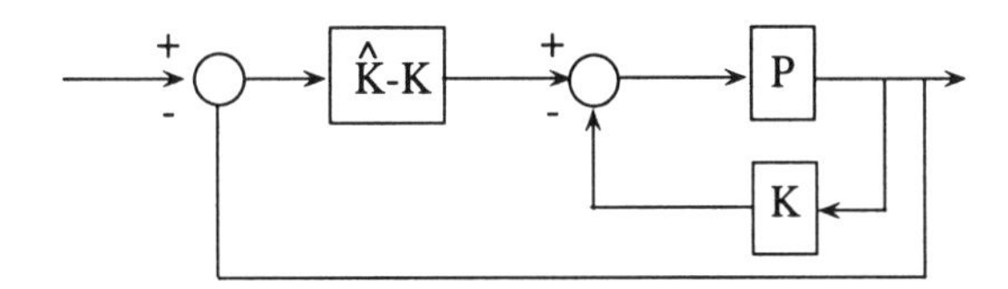

Figure 3.2.2. Redrawing of Figure 3.2.1

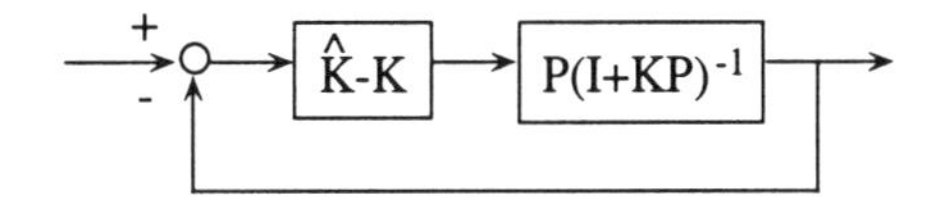

Figure 3.2.3. Redrawing of Figure 3.2.2

Using these redrawings, it can be concluded (Zhou, Doyle and Glover, 1996) that $\hat{K}(s)$ is a stabilizing compensator under the following sufficient two conditions:

1. $K(s)$ and $\hat{K}(s)$ have the same number of poles in the open right half plane, and $K(s) - \hat{K}(s)$ is bounded on the imaginary axis.

2. Either

$$\left\| (K - \hat{K}) P (I + KP)^{-1} \right\|_\infty < 1 \tag{3.2.1}$$

or

$$\left\| (I + PK)^{-1} P (K - \hat{K}) \right\|_\infty < 1. \tag{3.2.2}$$

This observation suggests a minimization problem: find a $\hat{K}(s)$ of prescribed degree satisfying Condition 1 and which minimizes either the left side of (3.2.1) or the left side of (3.2.2). It is evident that this problem is like the problems addressed in the first chapter, save that now a *frequency weighting* matrix, $P(I + KP)^{-1} = (I + PK)^{-1}P$ has been introduced.

The weight serves to say that it is more important to approximate $K(s)$ well at certain frequencies (where the weight is high) than at others. If we think of the scalar case to guide our intuition, we see that the weight will be small if either $|P|$ is small, or $|K|$ is large. It is likely to be greatest nearer the unity gain cross-over frequency for the loop gain PK—especially if there is no attempt to use K to increase the closed loop bandwidth beyond that of the open loop P. Evidently then, it is more important to approximate the controller accurately near the crossover frequency, an idea which is familiar from classical control.

In the unweighted approximation case, the problem of finding a minimizing low order approximation is generally too hard, and we rely on establishing methods which generally will bring us close to the minimum. The same turns out to be true in the weighted case. We shall actually present three broad classes of methods, one based on adjustment of the balanced truncation approach and due to Enns (1984), another based on generalization of the Hankel norm approximation approach and described in Latham and Anderson (1985) and Anderson (1986), and a third using a partial fraction expansion trick, which can then be combined with balanced truncation or Hankel norm reduction (Sreeram and Anderson, 1995; Zhou, 1993b; Zhou, 1995). This third method is an outgrowth of the weighted Hankel norm reduction procedure of Latham and Anderson (1985).

Closed-loop transfer function matrix consideration for frequency weighting

The closed-loop transfer function matrices with $K(s)$ and $\hat{K}(s)$ are

$$W(s) = P(s)K(s)\left[I + P(s)K(s)\right]^{-1}, \tag{3.2.3}$$

$$\hat{W}(s) = P(s)\hat{K}(s)\left[I + P(s)\hat{K}(s)\right]^{-1}. \tag{3.2.4}$$

Approximately, there holds

$$\hat{W}(s) - W(s) = \left[I + P(s)K(s)\right]^{-1} P(s)\left[\hat{K}(s) - K(s)\right]\left[I + P(s)K(s)\right]^{-1}. \tag{3.2.5}$$

This suggests the following approximation problem. Find $\hat{K}(s)$ of nominated degree so that

1. $K(s)$ and $\hat{K}(s)$ have the same number of poles in the open right half plane, and $K(s) - \hat{K}(s)$ is bounded on the imaginary axis.

2. The index

$$J = \left\| V_1(s)\big[\hat{K}(s) - K(s)\big]V_2(s)\right\|_\infty \tag{3.2.6}$$

is minimized, where $V_1 = (I + PK)^{-1}P$ and $V_2 = (I + PK)^{-1}$.

Comparing the left side of (3.2.1) or (3.2.2) with the index of (3.2.6) shows that in the latter index, we now have two-sided frequency weighting, and there is reduced weighting placed on frequencies in the high loop-gain region. Of course, in the $SISO$ case, the two-sided weighting can be made one-sided.

Another measure of closed-loop transfer function closeness is provided by the relative error $W(s)^{-1}[\hat{W}(s) - W(s)]$. From (3.2.5) and assuming that $P(s)$ and $K(s)$ are square and invertible, the following approximate equality can be easily obtained:

$$W(s)^{-1}\big[\hat{W}(s) - W(s)\big] = K^{-1}(s)\big[\hat{K}(s) - K(s)\big]\big[I + P(s)K(s)\big]^{-1}. \tag{3.2.7}$$

Again, this shows two sided weighting—but of a special kind. For obvious reasons, we can regard the right side of (3.2.7) as providing a frequency weighted relative error, see Kim, Anderson and Madievski (1995b).

Controller input spectrum based consideration for frequency weighting

Imagine the original closed-loop system, with external reference inputs and additive external noise. The combination of all these signals will give rise to a signal at the input of the controller, call it $q(t)$, and if the external signals are all characterised in terms of power spectra, then it will be possible to derive the power spectrum of the input signal to the controller.

In order that an approximation $\hat{K}(s)$ of $K(s)$ be a good approximation, it would be reasonable to require that the approximation be most accurate in those frequency bands encountered in actual operation. Thus if $q(t)$ has little spectral energy in one band, $K(j\omega)$ need not be closely approximated by $\hat{K}(j\omega)$ in that band, while if the spectral energy in another band is high, approximation needs to be accurate at frequencies in that band. Let $\Phi_{qq}(j\omega)$ denote the power spectrum of $q(t)$, and let $V(j\omega)$ be a stable, minimum phase spectral factor of $\Phi_{qq}(j\omega)$. [Thus $VV^* = \Phi_{qq}(j\omega)$] Then the approximation problem becomes: Find $\hat{K}(s)$ of nominated degree so that

1. $K(s)$ and $\hat{K}(s)$ have the same number of poles in the open right half plane, and $K(s) - \hat{K}(s)$ is bounded on the imaginary axis.

2. $\|[K(s) - \hat{K}(s)]V(s)\|_\infty$ is minimized.

H_∞ design objective consideration for frequency weighting

In H_∞ control design problems, the task is to find a controller which stabilizes the closed loop and ensures that the H_∞ gain from a specified set of plant inputs to a specified set of plant outputs is less than some nominated figure (Green and Limebeer, 1995; Zhou *et al.*, 1996). The controller obtained by the standard theory usually has order equal to or roughly equal to that of the plant, and so there is a need for a controller reduction algorithm that will ensure retention of the gain constraint for the combination of the plant and reduced order controller, as well as closed-loop stability. Material of Goddard and Glover (1993), also presented in Zhou *et al.* (1996), explains how to find weights $V_1(s)$ and $V_2(s)$ such that if

1. $K(s)$ and $\hat{K}(s)$ have the same number of poles in the open right half plane, and $K(s) - \hat{K}(s)$ is bounded on the imaginary axis, and

2. $$\left\| V_1(s)\left[\hat{K}(s) - K(s)\right]V_2(s)\right\|_\infty < 1 \tag{3.2.8}$$

then $\hat{K}(s)$ is stabilizing and ensures satisfaction of the gain constraint. The calculation of V_1 and V_2 is long and will be omitted here; these weights of course depend on the plant being controlled. Later in this book, we shall indicate another approach to controller reduction for H_∞ design, and present more details at that point.

Further points on frequency weighting and controller reduction

In this subsection, we make a number of small points:

1. Other weights than those indicated above may be appropriate on occasion. For example if the spectrum of external inputs were known, that could appear in a weight in considering the closed-loop transfer function error.

2. As will be seen in the next chapter, by using another representation of $K(s)$ via coprime fractions, a whole new set of frequency weighted problems can be obtained, some of which have the same conceptual basis as those of this section.

3. There is no single best weight, and if one concentrates on stability, it may be that performance is poor, and conversely.

4. The approximation problems posed are not fully appropriate for controllers with unstable or imaginary axis poles. Consider for example a controller containing a pure integrator. The approximation problems introduced all demand that the approximating controller also contain a pure integrator with *precisely* the same residue. This seems unnecessarily severe.

Frequency weighting and plant reduction

We have earlier argued that it is preferable to contemplate controller reduction after design of a high order controller than plant reduction before design of any controller, especially a low order one. On some occasions, it may however be necessary to reduce a model of a plant before any controller can be designed, perhaps because of limitations in the controller design software. We shall discuss briefly how one can address this problem.

Consider the arrangement of Figure 3.2.4.

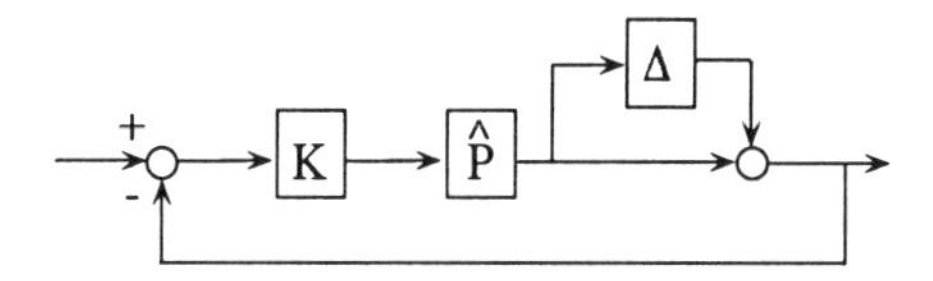

Figure 3.2.4. Closed loop with plant depicted as a multiplicative perturbation of model

The following result appears in Vidyasagar (1985), and is like results already obtained in connection with controller robustness.

Suppose that K stabilizes $\hat{P}$, that $\Delta = (P - \hat{P})\hat{P}^{-1}$ has no imaginary axis poles, and that P has the same number of poles in the closed right half plane as $\hat{P}$. If for all ω,

$$\left\| \Delta(j\omega)\, \hat{P}(j\omega)K(j\omega)\left[I + \hat{P}(j\omega)K(j\omega)\right]^{-1} \right\| < 1 \tag{3.2.9}$$

then K also stabilizes P.

This result indicates that if a controller K is designed to stabilize a nominal or reduced order model $\hat{P}$, satisfaction of (3.2.9) ensures that the controller will also stabilize the "true" plant P. Now in reducing a model of the plant, there will be concern not just to have this type of stability property, but also concern to have as little error as possible between the designed closed-loop system (based on $\hat{P}$) and the true closed-loop system (based on P). Extrapolation of the stability result then suggests that the goal should be not just to have (3.2.9), but to minimize the quantity on the left side of (3.2.9), or its greatest value as a function of frequency:

$$J = \left\| \Delta \hat{P} K\left[I + \hat{P}K\right]^{-1} \right\|_{\infty}. \tag{3.2.10}$$

Almost equivalently, we could try to minimize

$$\bar{J} = \max_{\omega}\left\{ \|\Delta(j\omega)\| \left\| \hat{P}(j\omega)K(j\omega)\left[I + \hat{P}(j\omega)K(j\omega)\right]^{-1} \right\| \right\}. \tag{3.2.11}$$

Now we have already pointed out a major difficulty with this sort of analysis: if we are reducing the plant without knowledge of the controller, we cannot calculate the measure because we do not know $K(s)$. Nevertheless, one could presume that, for a well designed system, $\hat{P}K(I+\hat{P}K)^{-1}$ will be close to I over the operating bandwidth of the system, and have norm smaller than 1 (tending to 0 as ω tends to ∞ in fact) outside the operating bandwidth of the system.

All this suggests that in the absence of knowledge of K, one should carry out a multiplicative error approximation by seeking to minimize $\|\Delta(j\omega)\|_\infty$. This is the prime rationale for unweighted multiplicative reduction of the plant. There is then a weighting function appropriate for reducing the plant P, namely P^{-1}. (Recall from our discussion of relative error approximation that the same approximant achieves the same bound for relative error as multiplicative error.)

If we know the cut-off frequency of the closed-loop system (prior to design of the controller), we can modify the above unweighted multiplicative criterion by a weighted one; the unweighted criterion seeks to have the multiplicative error uniformly small, including at frequencies well above the closed-loop cut-off frequency, where large multiplicative error could be tolerated with no risk of instability. Recognizing then that for some ω_0 it is likely that

$$\left\| \hat{P}(j\omega)K(j\omega)\left[I + \hat{P}(j\omega)K(j\omega)\right]^{-1} \right\| < \alpha\left(\omega^2 + \omega_0^2\right)^{-1/2}, \tag{3.2.12}$$

for some constant α, we see that there is motivation for an index such as

$$J = \left\| P^{-1}\left(P - \hat{P}\right)V \right\|, \tag{3.2.13}$$

where $V(s) = (s + \omega_0)^{-1}I$ provides the weighting. In this way, there will be more accurate multiplicative approximation in the closed-loop passband.

Main points of the section

1. Controller reduction can be formulated as a frequency-weighted error minimization problem.
2. The frequency weights can be derived by considering stability robustness, closed-loop transfer function error, controller input spectrum or H_∞ design objectives.
3. Open-loop unstable controllers are not well treated by the theory.
4. Plant reduction can be considered as a relative error problem together with a possible frequency weighting involving a 20 dB per decade frequency roll-off.

3.3 Frequency Weighted Balanced Truncation

We shall begin this section by describing an algorithm generalizing the balanced truncation algorithm and allowing introduction of one-sided frequency weighting. As with the balanced truncation algorithm, one can describe it using a series of transformations which include the numerically undesirable step of actually computing a balanced realization; then one can also find a variant on the algorithm which avoids this task, but at the same time is less transparent. We will consider issues of stability of the reduced order entity, and the calculation of error bounds, before tackling two-sided weighting.

Generalizing balanced truncation

Suppose our task is to find a low order $\hat{K}(s)$ with a prescribed number of poles which approximates a high order $K(s)$ under the condition that $K(s)$ and $\hat{K}(s)$ have the same number of poles in the open right half plane, and $K(s) - \hat{K}(s)$ is bounded on the imaginary axis; the approximation is in the sense of having the index

$$J = \left\| \left(K(s) - \hat{K}(s)\right) V(s) \right\|_\infty \tag{3.3.1}$$

small, even if not minimal. In this index, the weight $V(s)$ is assumed to have all its poles in the open left half plane. (It can be verified that the various weightings introduced in the last section have this property.)

It makes sense at once to introduce a simplification: suppose that with $K(s)$ represented as the sum of a transfer function matrix with all poles in the open left half plane, (the stable part of $K(s)$), and a transfer function matrix with all poles in the closed right half plane, (the unstable part of $K(s)$), the unstable part is simply copied into $\hat{K}(s)$, and the stable part of $\hat{K}(s)$ is determined by approximating the stable part of $K(s)$, taking no account of the unstable part of $K(s)$. In effect, we are reducing the approximation problem to one involving stable $K(s)$ only. Further, if $K(s)$ has a direct feedthrough term, we can copy that into $\hat{K}(s)$. So we can assume without loss of generality that $K(s)$ is also strictly proper.

Now suppose that state-variable realizations of $K(s)$ and $V(s)$ are given by

$$K(s) = \left[\begin{array}{c|c} A & B \\ \hline C & 0 \end{array}\right], \qquad V(s) = \left[\begin{array}{c|c} A_v & B_v \\ \hline C_v & D_v \end{array}\right]. \tag{3.3.2}$$

It is trivial to see that a state-variable realization of the product $K(s)V(s)$ is given by

$$K(s)V(s) = \left[\begin{array}{cc|c} A & BC_v & BD_v \\ 0 & A_v & B_v \\ \hline C & 0 & 0 \end{array}\right] = \left[\begin{array}{c|c} \bar{A} & \bar{B} \\ \hline \bar{C} & 0 \end{array}\right]. \tag{3.3.3}$$

The idea of frequency-weighted balanced truncation is to balance and then truncate those parts of the system or its state vector which are clearly associated with $K(s)$

rather than $V(s)$. Accordingly, let $\bar{P}$ be the controllability grammian associated with (3.3.3):

$$\bar{P}\bar{A}^T + \bar{A}\bar{P} + \bar{B}\bar{B}^T = 0 \tag{3.3.4}$$

and suppose that the top left block of $\bar{P}$ is denoted simply by P. Suppose also that $\bar{Q}$ is the observability grammian associated with (3.3.3) and denote its top left corner block by Q. It is not hard to see that, as a result of the triangular structure of $\bar{A}$ and $\bar{C}$, Q is given by the equation

$$QA + A^T Q + C^T C = 0. \tag{3.3.5}$$

We now suppose a coordinate basis change is made to $[A, B, C]$ which ensures that after the change, there holds $P = Q = \text{diag}[\lambda_1, \lambda_2, \ldots, \lambda_n]$, with $\lambda_i \geq \lambda_{i+1}$. This coordinate basis change is the balancing step, and it is achievable in just the same manner as in the unweighted case; the only difference is that the controllability matrix being used, *viz.* P, has been calculated differently. Notice that there is no change to A_v, B_v, C_v as a result of this coordinate basis change. We call the new realization a frequency-weighted balanced realization, and the quantities λ_i can be called (albeit with limited justification) frequency weighted Hankel singular values. Now, as before, the frequency weighted approximation is obtained by eliminating rows and columns of A, B, C corresponding to the smallest λ_i, *i.e.*, in turn λ_n, λ_{n-1}, *etc.* Also as before, if there are repeated frequency weighted Hankel singular values, either all repetitions of a particular value are thrown away or all are retained.

The greatest computational burden is associated with computing the controllability grammian. If the plant and controller each have order n, the weighting will have order $2n$ when the stability based weighting is being used, and apparently the grammian will involve a Lyapunov equation of dimension $3n$ by $3n$. However, there is some redundancy as it turns out, and one can get by with a Lyapunov equation of dimension $2n$ by $2n$ (from which an n by n submatrix P has to be selected).

The above procedure deals with input weighting; when there is output weighting instead of input weighting, the procedure is very similar. The computation of the controllability grammian now becomes straightforward, while the computation of the observability grammian must now reflect the presence of the weight.

The above algorithm description carries within it a balancing step. This can be avoided (on the grounds of its numerical undesirability) by using triangular transformed versions rather than diagonal transformed versions of the product PQ. Compute ordered Schur decompositions of PQ, with the eigenvalues of PQ in ascending and descending order:

$$V_A^T PQV_A = S_{\text{asc}}, \qquad V_D^T PQV_D = S_{\text{des}}, \tag{3.3.6}$$

where V_A and V_D are orthogonal and S_{asc} and S_{des} are upper triangular. Define submatrices as follows, with k the dimension of the reduced order system.

$$V_a = V_A \begin{bmatrix} 0 \\ I_k \end{bmatrix}, \qquad V_d = V_D \begin{bmatrix} I_k \\ 0 \end{bmatrix} \tag{3.3.7}$$

and determine a singular value decomposition of $V_a^T V_d$:

$$U_L S U_R^T = V_a^T V_d. \tag{3.3.8}$$

Now define transformation matrices:

$$S_L = V_a U_L S^{-1/2}, \qquad S_R = V_d U_R S^{-1/2}. \tag{3.3.9}$$

Notice that $S_L^T S_R = I$

The reduced order $\hat{K}(s)$ is now defined by

$$\hat{A} = S_L^T A S_R, \qquad \hat{B} = S_L^T B, \qquad \hat{C} = C S_R. \tag{3.3.10}$$

Stability and error bounds

The stability question is: if $K(s)$ has all poles in the open left half plane, does $\hat{K}(s)$, obtained by one-sided frequency weighted balanced truncation, also have this property? Enns (1984) established the answer in the affirmative. Focus on the version of the algorithm that actually computes diagonal P and Q. Recall that Q satisfies the Lyapunov equation (3.3.5). After truncation, a leading square block submatrix of Q (which is also diagonal)—call it $\hat{Q}$—will satisfy

$$\hat{Q}\hat{A} + \hat{A}^T Q + \hat{C}^T \hat{C} = 0. \tag{3.3.11}$$

Since $\hat{Q}$ is positive definite, it follows that the eigenvalues of $\hat{A}$ lie in the closed left half plane. It is moreover not hard to check that if $\hat{A}$ has a pure imaginary eigenvalue, the associated eigenvector lies in the nullspace of $\hat{C}$, *i.e.*, any imaginary axis eigenvalue of $\hat{A}$ corresponds to a nonobservable mode of the reduced order system. Accordingly, it can be discarded, and we conclude that the reduced order system transfer function, or its state-variable realization given by (3.3.10)—but with imaginary axis unobservable modes possibly having to be eliminated—has the desired stability property.

We turn now to a consideration of error bounds. Unfortunately, there is no neat *a priori* error bound as there is in the unweighted case. What is in fact possible (Kim, Anderson and Madievski, 1995a) is to bound the approximation error in terms of L_∞ norms of transfer function matrices of smaller order than the error itself. The error bound formula is as follows:

$$\left\| [K(s) - K_r(s)]\, V(s) \right\|_\infty \le 2 \sum_{r=k+1}^{n} \sqrt{\lambda_r^2 + \alpha_r \lambda_r^{3/2}}. \tag{3.3.12}$$

Here, the λ_r are the frequency weighted Hankel singular values defined above while the α_r are constants evaluated as follows

$$\alpha_r = \left\| S_r(s) \right\|_\infty \left\| Z_v(s) \right\|_\infty, \tag{3.3.13}$$

$$Z_v(s) = \text{causal part of } V(s)V^T(-s), \tag{3.3.14}$$

$$S_r(s) = A_{21}^{r-1}(sI - A_{r-1})^{-1} B_{r-1} + b_r, \tag{3.3.15}$$

where $A_n = A$, $B_n = B$, $C_n = C$ after balancing has occurred, and

$$A_r = \begin{bmatrix} A_{r-1} & A_{12}^{r-1} \\ A_{21}^{r-1} & a_{rr} \end{bmatrix}, \qquad B_r = \begin{bmatrix} B_{r-1} \\ b_r \end{bmatrix}. \tag{3.3.16}$$

(Thus A_r comprises the first r rows and columns of the balanced A, and B_r the first r rows of the balanced B.) Note that the calculation of $\|Z_v(s)\|_\infty$ is straightforward, and only has to be done once, *i.e.*, it is independent of r. The calculation of α_n however involves the evaluation of the L_∞ norm of a transfer function matrix of degree $n-1$. This may well be costly, though not necessarily as costly as evaluating the L_∞ norm of the error itself, the transfer function for which has degree $n + k + degV$.

Generalizing balanced truncation with two-sided weighting

In this subsection, we consider the problem of finding a low order approximation $\hat{K}(s)$ to a strictly proper $K(s)$ with all poles in the open left half plane which at the same time seeks to ensure that the following index is small:

$$J = \left\| V_1(s)\left[\hat{K}(s) - K(s)\right]V_2(s) \right\|_\infty. \tag{3.3.17}$$

We assume that both $V_1(s)$ and $V_2(s)$ have all poles in the open left half plane.

Suppose that state-variable realizations are available as

$$K(s) = \begin{bmatrix} A & B \\ C & 0 \end{bmatrix}, \qquad V_1(s) = \begin{bmatrix} A_1 & B_1 \\ C_1 & D_1 \end{bmatrix}, \qquad V_2(s) = \begin{bmatrix} A_2 & B_2 \\ C_2 & D_2 \end{bmatrix}. \tag{3.3.18}$$

It is trivial to see that the product $V_1(s)K(s)V_2(s)$ has a state-variable realization given by

$$V_1(s)K(s)V_2(s) = \left[\begin{array}{ccc|c} A & 0 & BC_2 & BD_2 \\ B_1C & A_1 & 0 & 0 \\ 0 & 0 & A_2 & B_2 \\ \hline D_1C & C_1 & 0 & 0 \end{array}\right] = \left[\begin{array}{c|c} \bar{A} & \bar{B} \\ \hline \bar{C} & 0 \end{array}\right]. \tag{3.3.19}$$

Once again, $\bar{P}$ and $\bar{Q}$ are defined as the controllability and observability grammians of the above state-variable realization, and P and Q are defined as the upper left $n \times n$ blocks of $\bar{P}$ and $\bar{Q}$, where the matrix A is $n \times n$. The idea is of course to modify just that part of the system (3.3.19) which corresponds to $K(s)$, *i.e.*, to modify in some way the first n entries of the state vector. This is done by treating P and Q in the same way as was done for unweighted balanced truncation, or the one-sided frequency weighted balanced truncation described a little earlier. Therefore one changes the coordinate basis so that after the change the transformed versions of P and Q become equal and diagonal, with reducing entries down the diagonal; then one truncates.

Alternatively, one can use the Schur form algorithm, which involves only orthogonal transformations, rather than general similarity transformations. We omit the details.

One small point is worth noting. The matrices P and Q have been described as being submatrices of $\bar{P}$ and $\bar{Q}$, which have dimension equal to the sum of the orders of $K(s)$, $V_1(s)$ and $V_2(s)$. Some saving is possible in the calculation of P and Q, which satisfy the following lower order equations, easily derived from the equations for $\bar{P}$ and $\bar{Q}$ by exploiting the special structure of $\bar{A}$, $\bar{B}$ and $\bar{C}$.

$$\begin{bmatrix} A & BC_2 \\ 0 & A_2 \end{bmatrix}\begin{bmatrix} P & P_{02} \\ P_{02}^T & P_{22} \end{bmatrix} + \begin{bmatrix} P & P_{02} \\ P_{02}^T & P_{22} \end{bmatrix}\begin{bmatrix} A^T & 0 \\ C_2^T B^T & A_2^T \end{bmatrix} + \begin{bmatrix} BD_2 \\ B_2 \end{bmatrix}\begin{bmatrix} BD_2 \\ B_2 \end{bmatrix}^T = 0, \tag{3.3.20}$$

$$\begin{bmatrix} Q & Q_{01} \\ Q_{01}^T & Q_{11} \end{bmatrix}\begin{bmatrix} A & 0 \\ B_1 C & A_1 \end{bmatrix} + \begin{bmatrix} A^T & C^T B_1^T \\ 0 & A_1^T \end{bmatrix}\begin{bmatrix} Q & Q_{01} \\ Q_{01}^T & Q_{11} \end{bmatrix} + \begin{bmatrix} C^T D_1^T \\ C_1^T \end{bmatrix}\begin{bmatrix} C^T D_1^T \\ C_1^T \end{bmatrix}^T = 0. \tag{3.3.21}$$

In contrast to the one-sided frequency weighting case, there is no guarantee that $\hat{K}(s)$ will be stable. For a number of years it was conjectured that stability was guaranteed; however, counter-examples have been constructed. The following remarkably simple example is taken from Sreeram, Anderson and Madievski (1995)

$$K(s) = \frac{8s^2 + 6s + 2}{s^3 + 4s^2 + 5s + 2}, \qquad V_1(s) = \frac{1}{s+4}, \qquad V_2(s) = \frac{1}{s+3}. \tag{3.3.22}$$

The diagonalized frequency weighted controllability and observability grammians after balancing are

$$P = Q = \text{diag}\,[0.0513, 0.0417, 0.0057] \tag{3.3.23}$$

and while the second order approximation is stable, the first order approximation is

$$\hat{K}(s) = -\frac{0.1563}{s - 0.1085}. \tag{3.3.24}$$

Not surprisingly, it is possible to obtain reduced order quantities with poles even lying on the imaginary axis. This makes the search for an error bound, especially an *a priori* error bound, almost certainly futile. The so-called bounds of Kim *et al.* (1995a) are effectively not *a priori—i.e.*, one must almost execute the whole reduction process in order to obtain the bound, and even then the issue of unstable poles can only be dealt with in an imperfect way.

The variant of Lin and Chiu and stability retention

In Lin and Chiu (1992) and as further developed in Sreeram *et al.* (1995), a minor adjustment is proposed to the scheme of Enns (1984) which guarantees stability for the reduced order $\hat{K}(s)$ with both one-sided and two-sided weighting. There is a logical motivation for the adjusted scheme, but experience reported in Sreeram *et al.* (1995) shows that on some occasions the actual error with the Enns scheme is lower, while on other occasions the actual error with the modified scheme is better. Therefore, there can be no recommendation that the modification always be used.

We shall present the modification for the two-sided weighting problem, the one-sided problem being after all simply a special case of the two-sided problem. Recall the equations used to determine the frequency-weighted grammians P and Q, (3.3.20) and (3.3.21) above. The modification consists in not balancing with respect to P and Q but with respect to two matrices derived by a Schur complementation operation. These are $\tilde{P}$ and $\tilde{Q}$ respectively, given by

$$\tilde{P} = P - P_{02}P_{22}^{-1}P_{02}^T, \qquad \tilde{Q} = Q - Q_{01}Q_{11}^{-1}Q_{01}^T. \tag{3.3.25}$$

In case there is only one-sided frequency weighting, only one of $\tilde{P}$ and $\tilde{Q}$ differs from P or Q. Although there is no benefit from a stability point of view in making the replacement, it may be that in the one-sided case, the replacement gives a lower error for a given order of $\hat{K}(s)$. Also, in the two-sided case, it is actually only necessary to replace one of P and Q with $\tilde{P}$ or $\tilde{Q}$ to guarantee stability of the reduced order $\hat{K}(s)$.

Let us first observe now why the replacement of both P and Q guarantees stability, and then examine why there is some *a priori* motivation (apart from the stability outcome) in making the replacement.

From (3.3.20) and the definition of $\tilde{P}$, one can show that

$$\begin{bmatrix} A & X_{12} \\ 0 & A_2 \end{bmatrix}\begin{bmatrix} \tilde{P} & 0 \\ 0 & P_{22} \end{bmatrix} + \begin{bmatrix} \tilde{P} & 0 \\ 0 & P_{22} \end{bmatrix}\begin{bmatrix} A^T & 0 \\ X_{12}^T & A_2^T \end{bmatrix} + \begin{bmatrix} X \\ B_2 \end{bmatrix}\begin{bmatrix} X \\ B_2 \end{bmatrix}^T = 0, \tag{3.3.26}$$

where

$$X_{12} = AP_{02}P_{22}^{-1} + BC_2 - P_{02}P_{22}^{-1}A_2, \qquad X = BD_2 - P_{02}P_{22}^{-1}B_2. \tag{3.3.27}$$

Similar equations hold starting with (3.3.21). Taking the 1-1 block entry of (3.3.26) and its companion equation involving $\tilde{Q}$ leads to

$$A\tilde{P} + \tilde{P}A^T + XX^T = 0, \qquad \tilde{Q}A + A^T\tilde{Q} + Y^TY = 0, \tag{3.3.28}$$

where $Y = D_1C - C_1Q_{11}^{-1}Q_{01}$. Now the frequency weighted reduction procedure for the original $K(s)$, but with the modification that P and Q are replaced by $\tilde{P}$ and $\tilde{Q}$, is equivalent to an unweighted balanced truncation for the state-variable triple $\{A, X, Y\}$. This triple is necessarily minimal if $\tilde{P}$ and $\tilde{Q}$ are positive definite. They in turn will be positive definite if the solutions of (3.3.20) and (3.3.21) have this property;

absence of stable pole-zero cancellations between the weights and $K(s)$ is necessary and sufficient for this; we shall assume here that no such cancellations occur.

Since unweighted balanced truncation always preserves stability (assuming that when Hankel singular values are repeated, all repeats are either included or excluded from the reduced order model), so must the reduced order $\hat{K}(s)$ (which has the same $\hat{A}$ matrix as the reduced order triple $\{\hat{A}, \hat{X}, \hat{Y}\}$) have the stability property.

To motivate the procedure, recall first an aspect of unweighted balanced truncation for a system with state-variable realization $\{A, B, C\}$, where all eigenvalues of A lie in the open left half plane. Consider the optimization problem of minimizing the input energy $\int_{-T}^{0} u^T(t)u(t)dt$ to the system

$$\dot{x} = Ax + Bu, \tag{3.3.29}$$

under the constraint that $x(-T) = 0$ and that a certain $x(0) = x_0$ is attained at time 0. When T approaches ∞, the minimum energy is

$$E_{\min} = x_0^T P^{-1} x_0, \tag{3.3.30}$$

where P is the controllability grammian. Also, consider the unforced system

$$\dot{x} = Ax, \qquad z = Cx + v, \tag{3.3.31}$$

where $v(t)$ denotes zero mean unit variance gaussian white noise. The mean square error covariance in estimating the initial state $x(0)$ from measurements $z(t)$ for $t > 0$ is Q^{-1}. Balanced realizations are those where the minimum energy weighting matrix equals the error covariance matrix.

Now consider a system with input weighting. In place of (3.3.29), we have

$$\begin{bmatrix} \dot{x} \\ \dot{x}_2 \end{bmatrix} = \begin{bmatrix} A & BC_2 \\ 0 & A_2 \end{bmatrix} \begin{bmatrix} x \\ x_2 \end{bmatrix} + \begin{bmatrix} BD_2 \\ B_2 \end{bmatrix} u. \tag{3.3.32}$$

The minimum energy in arriving at $[x^T(0)\ x_2^T(0)]^T$ from the zero state in the infinitely remote past is

$$E_{\min} = \begin{bmatrix} x^T(0) & x_2^T(0) \end{bmatrix} \begin{bmatrix} P & P_{02} \\ P_{02}^T & P_{22} \end{bmatrix}^{-1} \begin{bmatrix} x(0) \\ x_2(0) \end{bmatrix}. \tag{3.3.33}$$

If we insist that $x_2(0)$ be zero, this minimum energy is precisely $x^T(0)\tilde{P}^{-1}x(0)$. If with $x(0)$ fixed we allow $x_2(0)$ to be chosen so as to minimize the minimum energy, not just with respect to $u(.)$ but also with respect to $x_2(0)$, then the minimum energy is $x^T(0)P^{-1}x(0)$. Obviously, in seeking to generalize the calculations applicable to unweighted balanced truncation, there is a case for both alternatives. A similar argument applies in considering the choice between Q and $\tilde{Q}$, see Sreeram *et al.* (1995).

These remarks are not intended to assert that one choice is inherently better than the other. We re-emphasize that in actual examples, sometimes one choice is superior and sometimes another choice is superior in terms of actual error.

The connection with relative error approximation

Let G be square and nonsingular, with G and G^{-1} having all poles in the open left half plane. The problem of minimizing the index

$$J = \left\| G^{-1}(s)\left[G(s) - \hat{G}(s)\right] \right\|_{\infty} \tag{3.3.34}$$

was studied in the previous chapter, and is regarded as a relative error problem. Provided G^{-1} is stable and proper, it is also a frequency weighted reduction problem in which the left hand weight happens to be special in terms of its relation to the object being reduced. In Zhou (1995) and Zhou *et al.* (1996) there is a comparison of the use of the two reduction methods, balanced stochastic truncation and frequency weighted balanced truncation, for this index, under the assumption that G and G^{-1} are stable and proper. It turns out that both methods give the same reduced order $\hat{G}_k(s)$ of order k, and that the Hankel singular values of the transfer function matrix we termed $F_c(s)$ in the discussion of the balanced stochastic truncation method and the frequency weighted Hankel singular values arising in frequency weighted balanced truncation are relatable by a fairly simple formula. Also, when the controllability grammians in the two problems are made equal (which turns out to be easy) the relevant observability grammians can be related.

In more detail, the weighted grammians P and Q for the frequency weighted balanced truncation problem for

$$G(s) = \left[\begin{array}{c|c} A & B \\ \hline C & D \end{array}\right], \tag{3.3.35}$$

(with $G^{-1}(s)$ stable and proper) are obtained as the solutions of

$$PA^T + AP + BB^T = 0, \tag{3.3.36}$$

$$Q(A - BD^{-1}C) + (A - BD^{-1}C)^T Q + C^T D^{-T} D^{-1} C = 0, \tag{3.3.37}$$

with the observability grammian of

$$G^{-1}(G - D) = \left[\begin{array}{cc|c} A & 0 & B \\ -BD^{-1}C & A - BD^{-1}C & 0 \\ \hline D^{-1}C & D^{-1}C & 0 \end{array}\right] \tag{3.3.38}$$

given by

$$\bar{Q} = \begin{bmatrix} Q & Q \\ Q & Q \end{bmatrix}. \tag{3.3.39}$$

Suppose that the associated weighted Hankel singular values are σ_i, *i.e.*, if P and Q are balanced then

$$P = Q = \mathrm{diag}\,[\sigma_1, \sigma_2, \ldots, \sigma_n], \tag{3.3.40}$$

with $\sigma_1 \geq \sigma_2 \geq \cdots \geq \sigma_n$. Denote by $\sigma_{1m}, \sigma_{2m}, \ldots, \sigma_{nm}$ the Hankel singular values of the causal part of $H_*^{-1}G$, where $GG_* = H_*H$ and H is stable and stably invertible.

The σ_{im} are the square roots of the eigenvalue of PQ_m, where, in accordance with Section 2.2, Q_m is the stabilizing solution of

$$Q_mA + A^TQ_m + \left[C - (CP + DB^T)Q_m\right]^T(DD^T)^{-1}\left[C - (CP + DB^T)Q_m\right] = 0. \tag{3.3.41}$$

It is possible to show that $Q_m = Q(I + PQ)^{-1}$ and that

$$\sigma_{im} = \frac{\sigma_i}{\sqrt{1 + \sigma_i^2}}. \tag{3.3.42}$$

Let $\nu_1 = \sigma_{i_1m}, \nu_2 = \sigma_{i_2m}, \ldots \nu_N = \sigma_{i_Nm}$ denote the distinct values of σ_{im}, in descending order. Then the error bound

$$\left\|G^{-1}(G - \hat{G})\right\|_\infty \leq \prod_{j=r+1}^{N} \frac{1 + \nu_j}{1 - \nu_j} - 1 \tag{3.3.43}$$

is equivalently

$$\left\|G^{-1}(G - \hat{G})\right\|_\infty \leq \prod_{j=r+1}^{N} \left(1 + 2\sigma_{i_j}\sqrt{1 + \sigma_{i_j}^2} + 2\sigma_{i_j}^2\right) - 1. \tag{3.3.44}$$

This problem, regarded as a frequency weighted balanced truncation problem, is therefore one for which a simple error bound formula can be found. For general frequency weighted balanced truncation problems, there is no known simple error bound formula.

Main points of the section

1. The balanced truncation procedure can be adapted to allow for frequency weighting.
2. With one-sided frequency weighting, frequency-weighted balanced truncation approximates a stable entity with another stable entity. This is not so for two-sided weighting. There is no attractive error bound formula which is known.
3. A variant on the frequency-weighted balanced truncation procedure is available which always assures stability (even for two-sided weighting), but may or may not give better error characteristics.

4. Relative error approximation via balanced stochastic truncation of a stable minimum phase plant can be viewed as a form of frequency weighted balanced truncation, and error bounds are available in this case.

3.4 Frequency Weighted Hankel Norm Reduction

Our task in this section is to provide a frequency-weighted reduction procedure using Hankel norm approximation ideas, as opposed to an extension of balanced truncation. A method was first proposed in Latham and Anderson (1985), with error bounds in Anderson (1986). Extensions can be found in Zhou (1993b) and Zhou (1995). Textbook treatments, with some extensions, can be found in Green and Limebeer (1995) and Zhou *et al.* (1996). Special results applicable to the case where the weight is a first order stable transfer function can be found in Hung and Glover (1985), with a multivariable version in Zhou (1993a).

We start with the index

$$J = \left\| V_1(s)\left[K(s) - \hat{K}(s)\right]V_2(s)\right\|_\infty \tag{3.4.1}$$

in which the weights V_1 and V_2 are square, and generally have all poles in the open left half plane. We shall make the additional assumption that their determinants are nonzero on the imaginary axis. As before, $K(s)$ is also restricted to having poles in the open left half plane, and we seek a lower order $\hat{K}$ with the same property. Now let W_1 and W_2 be defined by

$$V_{1*}(s)V_1(s) = W_{1*}(s)W_1(s), \qquad V_2(s)V_{2*}(s) = W_2(s)W_{2*}(s), \tag{3.4.2}$$

together with the requirement that these W_1 and W_2 together with their inverses have all their poles in the open *right* half plane. In the scalar case, it is easy to find W_1 and W_2, knowing the poles and zeros of V_1 and V_2. In general however, Riccati equations can be used to find W_1 and W_2, although if V_1 and V_2 are singular at $s = \infty$, the calculations are less straightforward (Zhou *et al.*, 1996).

Now since $W_1V_1^{-1}$ and $V_2^{-1}W_2$ are unitary matrices on the imaginary axis on account of (3.4.2), it follows that

$$J = \left\| W_1(s)\left[K(s) - \hat{K}(s)\right]W_2(s)\right\|_\infty. \tag{3.4.3}$$

Let $\sigma_1, \sigma_2, \ldots$ denote in descending order the Hankel singular values of the stable part of W_1KW_2. Then if $\hat{K}(s)$ is stable and has degree k, an immediate lower bound on J is

$$J \geq \sigma_{k+1}\left([W_1KW_2]_+\right). \tag{3.4.4}$$

To see this, observe that

$$\sigma_{k+1} = \inf_{Q \in RH_\infty^-(k)} \left\| W_1 K W_2 - Q \right\|_\infty \tag{3.4.5}$$

$$= \inf_{\bar{Q} \in RH_\infty^-(k)} \left\| W_1 (K - \bar{Q}) W_2 \right\|_\infty \tag{3.4.6}$$

$$\leq \inf_{\hat{K}} \left\| W_1 (K - \hat{K}) W_2 \right\|_\infty, \tag{3.4.7}$$

where $\hat{K} \in RH_\infty^-(k)$ and is stable. Here we have used the easily checked fact that $\bar{Q} \in RH_\infty^-(k)$ if and only if there holds $W_1 \bar{Q} W_2 \in RH_\infty^-(k)$.

The above calculations also provide the clue to the approximation procedure. Let Q solve the optimization problem (3.4.5). Define $\bar{Q}$ as

$$\bar{Q} = W_1^{-1} Q W_2^{-1} \tag{3.4.8}$$

so that $\bar{Q}$ satisfies the optimization problem (3.4.6). Finally, in a move reminiscent of unweighted Hankel norm approximation, set

$$\hat{K} = [\bar{Q}]_+ = \left[W_1^{-1} [Q]_+ W_2^{-1} \right]_+ . \tag{3.4.9}$$

Notice that $\hat{K}$ is the actual minimizer of the index $\| W_1 (K - \hat{K}) W_2 \|_H$ over all stable $\hat{K}$ of degree up to k. Note also that the second equation of (3.4.9) allows calculation of $\hat{K}(s)$ from the stable Hankel norm approximation of degree k of $W_1 K W_2$, *i.e.*, it is not necessary to find Q but enough to find just its stable part $[Q]_+$.

The final step of the approximation process involves the calculation implicit in (3.4.9). In Zhou *et al.* (1996), formulas are given for passing from a state-variable realization of each of W_1, W_2 and $[Q]_+$ to $\hat{K}$ using the second formula of (3.4.9). Assume that $W_1(\infty)$ and $W_2(\infty)$ are both nonsingular, and suppose the relevant state-variable realizations are as follows:

$$[Q]_+ = \begin{bmatrix} A_Q & B_Q \\ C_Q & D_Q \end{bmatrix}, \qquad W_1 = \begin{bmatrix} A_1 & B_1 \\ C_1 & D_1 \end{bmatrix}, \qquad W_2 = \begin{bmatrix} A_2 & B_2 \\ C_2 & D_2 \end{bmatrix}. \tag{3.4.10}$$

Then the constituent matrices of a state-variable realization of $\hat{K}(s)$, apart from the direct feedthrough term $\hat{K}(\infty)$, are given by solving

$$\hat{A} = A_Q, \tag{3.4.11}$$

$$\hat{B} = (B_Q - R B_2) D_2^{-1}, \tag{3.4.12}$$

where $R(A_2 - B_2 D_2^{-1} C_2) - A_Q R + B_Q D_2^{-1} C_2 = 0$ and

$$\hat{C} = D_1^{-1} \left[C_Q - C_1 S \right], \tag{3.4.13}$$

where $(A_1 - B_1 D_1^{-1} C_1)S - SA_Q + B_1 D_1^{-1} C_Q = 0.$

Note that both R and S are guaranteed to exist. This is because all eigenvalues of $(A_2 - B_2 D_2^{-1} C_2)$, $-A_Q$ and $(A_1 - B_1 D_1^{-1} C_1)$ lie in $\text{Re}[s] > 0$.

In general, *a priori* overbounds are not available for the error in the above algorithm. For the discrete-time version of the algorithm, rather complicated bounds can be found, see Anderson (1986). While these are L_∞ norms, they involve also the L_2 norm of the weight. Under bilinear transformation, the latter norm is not preserved, and this explains why a continuous-time result cannot be obtained from the discrete-time result.

For special weights, continuous-time bounds are however available. In Hung and Glover (1985) the problem is considered of finding a stable scalar $\hat{K}(s)$ of prescribed order to minimize the index

$$J = \left\| \frac{s+b}{s+a} \left[K(s) - \hat{K}(s) \right] \right\|_\infty , \tag{3.4.14}$$

where, without loss of generality, we can assume that a and b are positive. Denote by $\sigma_1, \sigma_2, \ldots$ the Hankel singular values in descending order of the stable part of $(s-a)^{-1}(s-b)K(s)$. By reducing one step at a time (using the algorithm above for each step) and by choosing the constant term in $\hat{K}(s)$ optimally, the following error bound results when $\hat{K}(s)$ has degree k:

$$\left\| \frac{s+b}{s+a} \left[K(s) - \hat{K}(s) \right] \right\|_\infty \le \left(1 + \frac{|a-b|}{a+b} \right) (\sigma_{k+1} + \cdots + \sigma_n). \tag{3.4.15}$$

In Zhou (1993a), a type of multivariable generalization of this bound is obtained. Suppose that $K(s)$ is m by m, that the right hand weight $V_2(s)$ is I, and the left hand weight $V_1(s)$ has state variable realization $A_1, B_1, C_1, D_1 = 0$ where B_1 is square and nonsingular—thus the weight has order equal to its dimension as a transfer function matrix. Then the bound is

$$\left\| V_1(s) \left[K(s) - \hat{K}(s) \right] \right\|_\infty \le 2(\sigma_{k+1} + \cdots + \sigma_n). \tag{3.4.16}$$

Once again, one has to select $\hat{D}$ in a particular way to achieve the bound.

Main points of the section

1. A frequency weighted Hankel norm approximation procedure is available.

2. A lower bound error formula is available.

3. For scalar weightings of first order, an upper bound error formula is available. A limited multivariable generalization of the bound is available.

3.5 Frequency Weighted Reduction Using Partial Fractions

In this section, we aim to explore another approach to frequency weighted reduction which is an outgrowth of the approach originally advanced in Latham and Anderson (1985) for weighted Hankel norm approximation; that approach implicitly appealed to certain properties of partial fraction expansions. We make the idea explicit here, and open the possibility of combining it with balanced truncation and not just Hankel norm approximation. The main reference is Sreeram and Anderson (1995).

As before, let $K(s)$ be the stable transfer function matrix which is to be reduced, and let $V_1(s)$ and $V_2(s)$, either of which may be absent in the sense of being replaced by the identity matrix, be weighting matrices. The basic idea of the scheme is as follows. If $K(s)$ has no pole in common with either of V_1 or V_2, then using a partial fraction expansion we can write

$$V_1 K V_2 = Z + \text{terms involving poles of } V_1 \text{ and } V_2 \text{ but not } K. \tag{3.5.1}$$

The poles of Z are all poles of K.

Next, Z is reduced by balanced truncation or Hankel norm reduction to $\hat{Z}$ say. Finally we find $\hat{K}$, with poles identical to those of $\hat{Z}$, such that

$$V_1 \hat{K} V_2 = \hat{Z} + \text{terms involving poles of } V_1 \text{ and } V_2 \text{ but not } \hat{K}. \tag{3.5.2}$$

This is not a wholly trivial task, and we shall explain how to do it in state variable terms. What can go wrong in this last step is that the weights can have a zero at the poles of $\hat{Z}$.

In case V_1 and V_2 have all poles and zeros in the open right half plane, as was demanded in the case of frequency weighted Hankel norm reduction, the decomposition (3.5.1) and the computation of $\hat{K}(s)$ required for (3.5.2) are both straightforward. If the reduction procedure on Z is Hankel norm reduction, *then the whole algorithm is indeed equivalent to the earlier presented frequency weighted Hankel norm reduction.*

We shall now indicate what is involved in state-variable type calculations for (3.5.1) and (3.5.2).

Suppose that minimal state-variable realizations are available as

$$K(s) = \left[\begin{array}{c|c} A & B \\ \hline C & 0 \end{array}\right], \qquad V_1(s) = \left[\begin{array}{c|c} A_1 & B_1 \\ \hline C_1 & D_1 \end{array}\right], \qquad V_2(s) = \left[\begin{array}{c|c} A_2 & B_2 \\ \hline C_2 & D_2 \end{array}\right] \tag{3.5.3}$$

and assume that $K(s)$ has no pole in common with either $V_1(s)$ or $V_2(s)$. Define X and Y as the solutions of

$$XA - A_1 X + B_1 C = 0, \qquad AY - YA_2 + BC_2 = 0. \tag{3.5.4}$$

(The condition on absence of common poles guarantees the existence of X and Y.) Then a state-variable realization for $Z(s)$ is given by

$$Z(s) = \left[\begin{array}{c|c} A & -YB_2 + BD_2 \\ \hline -C_1X + D_1C & 0 \end{array}\right]. \tag{3.5.5}$$

If there are common poles, either (3.5.4) is not solvable, or there are an infinite number of solutions; either case is too complicated to consider here. The realization of $Z(s)$ is not guaranteed to be minimal; indeed, if there is a zero of one of the weights cancelling a pole of $K(s)$, one can expect the realization to be nonminimal. If for example $V_1 = I$ and $V_2 = P(I + KP)^{-1}$, all poles of K are cancelled, and the method is useless. However, if $V_2(s)$ is replaced by $W_2(s)$, see (3.4.2), the method works.

Now let us turn to the determination of $\hat{K}(s)$ from $\hat{Z}(s)$, see (3.5.2). It is immediately clear that a helpful, possibly necessary, condition is that V_1 and V_2 be invertible. Then we would have

$$\hat{K}(s) = V_1^{-1}(s)\big[\hat{Z}(s) + \text{terms involving poles of } V_1 \text{ and } V_2\big]V_2^{-1}(s). \tag{3.5.6}$$

More precisely, noting the poles of $\hat{K}(s)$ should be those of $\hat{Z}(s)$, it would seem that if

$$\hat{Z}(s) = \sum_i \frac{Z_i}{s - a_i} \tag{3.5.7}$$

we should have

$$\hat{K}(s) = \sum_i V_1^{-1}(a_i)\frac{Z_i}{s - a_i}V_2^{-1}(a_i). \tag{3.5.8}$$

Thus $V_1(a_i)$ and $V_2(a_i)$ should be invertible. If V_1 and V_2 are not square, we cannot expect $\hat{K}$ to be unique; we need however V_1 to have full row rank and V_2 to have full column rank at the poles of $\hat{Z}$. The details of the calculation of $\hat{K}(s)$ are as follows. Suppose that

$$\hat{Z}(s) = \left[\begin{array}{c|c} \hat{A} & \hat{B}_z \\ \hline \hat{C}_z & 0 \end{array}\right]. \tag{3.5.9}$$

Suppose further that for all $\lambda = \lambda_i(\hat{A})$, the matrices

$$\begin{bmatrix} A_1 - \lambda I & B_1 \\ C_1 & D_1 \end{bmatrix}, \qquad \begin{bmatrix} A_2 - \lambda I & B_2 \\ C_2 & D_2 \end{bmatrix},$$

have full row rank and full column rank respectively. Then there exist $\hat{X}$, $\hat{Y}$, $\hat{B}$ and $\hat{C}$, unique if V_1 and V_2 are square, and satisfying the linear equations

$$\begin{aligned} \hat{X}\hat{A} - A_1\hat{X} + B_1\hat{C} = 0, \qquad & -C_1\hat{X} + D_1\hat{C} = \hat{C}_z, \\ \hat{A}\hat{Y} - \hat{Y}A_2 + \hat{B}C_2 = 0, \qquad & -\hat{Y}B_2 + \hat{B}D_2 = \hat{B}_z. \end{aligned} \tag{3.5.10}$$

The triple $\{\hat{A}, \hat{B}, \hat{C}\}$ determines a $\hat{K}$ such that (3.5.2) holds.

In the previous subsection, we recalled a special error bound result available for frequency weighted Hankel norm reduction when the weighting function was of degree 1 with $K(s)$ scalar, there being a generalization of the result for the case of square non-scalar $K(s)$. So also if one uses the above algorithm just presented with the same weighting function restriction, and with balanced truncation for the actual reduction step, a nice error bound result is available. Without having presented an algorithm in the form described above, Al-Saggaf and Franklin (1986) and Al-Saggaf and Franklin (1988) have dealt with this problem.

To give a high level view of their contribution from the viewpoint of the above algorithm, let us suppose that K is m by m, that V_2 is the identity, and V_1 is a full rank square strictly proper transfer function matrix of the form $C_1(sI - A_1)^{-1}B_1$ with B_1 square. Thus if $m = 1$, V_1 is a first order weighting function. From equations (3.5.1) and (3.5.2) it is evident that

$$V_1[K - \hat{K}] = [Z - \hat{Z}] + \text{terms involving poles of } V_1. \tag{3.5.11}$$

Now $\hat{K}$ has freedom in it to the point of being able to choose the direct feedthrough term $\hat{D}$. Hence we can rewrite as

$$V_1[K - \hat{K}] = [Z - \hat{Z}] + (\text{terms involving poles of } V_1) + V_1\hat{D}. \tag{3.5.12}$$

The degree restriction on V_1 actually means that it is possible to choose the constant matrix $\hat{D}$ so that the last two summands in (3.5.12) cancel, resulting in

$$V_1[K - \hat{K}] = [Z - \hat{Z}]. \tag{3.5.13}$$

(This is trivial to see in the scalar case, but requires several lines of proof in the matrix case.) As a result, the standard balanced truncation error bound for $Z - \hat{Z}$ provides a bound for the frequency weighted error also.

Main points of the section

1. The frequency weighted Hankel norm approximation procedure rests on relating parts of the partial fraction expansion of a transfer function with parts of the partial fraction expansion of a frequency weighted version of that transfer function.

2. This observation suggests another reduction procedure involving partial fractions and balanced truncation. For scalar systems with first order weights, an error bound formula is available. A limited matrix generalization of the bound can be found.

3.6 Multiplicative Approximation with Frequency Weighting

A problem of multiplicative approximation with frequency weighting is of the following type. The data are a high order object $G(s)$ and a weight $V(s)$. What is desired is a low order approximation $\hat{G}(s)$ to $G(s)$ such that the following index is small:

$$J = \left\| G^{-1}(G - \hat{G})V \right\|_{\infty}. \tag{3.6.1}$$

We shall restrict attention to the case where $G(s)$ is square, full rank and with all poles in the open left half plane. This is restrictive, but it allows us to apply a modification of balanced stochastic truncation, which normally requires a stable $G(s)$. Also, we shall assume that $V(s)$ is stable—this last without loss of generality if we *a priori* rule out purely imaginary poles. We have already given two instances of this problem arising—in controller approximation designed to keep small the relative error in the closed-loop transfer function matrix, and in plant approximation where the controller is unknown, but the desired closed-loop bandwidth may be known. Yet another example arises in the area of digital filtering. Often, an initial digital filter design results in a high order finite impulse response (FIR) filter; a low order infinite impulse response (IIR) filter is desired as an approximation, and usually in the passband the quality of approximation is measured by the dB and phase error, *i.e.*, by a relative error. On the other hand, in the stopband, the relative error ceases to be very important when the amplitude response is small in absolute terms; hence when the bandwidth of the filter is known, a weighting function providing decay beyond the bandwidth may be introduced.

In the case of plant approximation, it is to be expected that $G(s)$ will not be nonsingular on the extended $j\omega$ axis, but almost certainly zero at $s = \infty$. This causes a problem—more in the nature of an irritation—in a multiplicative approximation problem where G^{-1} appears in the index. It can be dealt with through use of the bilinear transformation

$$\begin{aligned} \tilde{G}(s) &= G\left(\frac{s-a}{-bs+1}\right), \\ \tilde{V}(s) &= V\left(\frac{s-a}{-bs+1}\right), \end{aligned} \tag{3.6.2}$$

where a and b are both small with $0 < a < b^{-1}$ (Safonov, 1987). Approximation occurs using the transformed quantities (in which there are no $j\omega$ axis problems, neither at infinity nor anywhere else) to find a $\hat{\tilde{G}}(s)$, and then the inverse transform is applied to obtain $\hat{G}(s)$ as

$$\hat{G}(s) = \hat{\tilde{G}}\left(\frac{s+a}{bs+1}\right). \tag{3.6.3}$$

In the remainder of this section, we shall assume that where needed such an approach is invoked.

The algorithm, due to Kim *et al.* (1995b), begins from a state-variable description of $G(s)$, say

$$G(s) = \left[\begin{array}{c|c} A & B \\ \hline C & D \end{array}\right] \tag{3.6.4}$$

and finds, just as in unweighted balanced stochastic truncation, matrices K, L, D_z and D_h such that

$$\begin{bmatrix} Z(s) & G(s) \\ H(s) & F_c(s) \end{bmatrix} = \begin{bmatrix} D_z & D \\ D_h & 0 \end{bmatrix} + \begin{bmatrix} C \\ K \end{bmatrix} (sI - A)^{-1} \begin{bmatrix} L & B \end{bmatrix}, \tag{3.6.5}$$

with

$$GG_* = H_* H = Z + Z_*. \tag{3.6.6}$$

Also, $H(s)$ is minimum phase. In normal balanced stochastic truncation, there is now a reduction of $F_c(s)$ using ordinary balanced truncation ideas, and the reduction induces a corresponding reduction in each of $G(s)$, $Z(s)$ and $H(s)$, although our focus is on the reduction of $G(s)$. Now for the weighted multiplicative reduction problem, one simply performs the frequency-weighted balanced truncation algorithm on the product $F_c(s)V(s)$, to find $\hat{F}_c(s)$ with $\|(F_c - \hat{F}_c)V\|_\infty$ small. This is the only change, and it is of course very straightforward to implement.

In unweighted balanced stochastic truncation, the right half plane zeros of $G(s)$ are copied into $\hat{G}(s)$. This does not happen in the weighted version, although Kim *et al.* (1995b) indicates that the number of right half plane zeros is preserved in all examples studied. The examples also clearly display the effect of the weight as compared with the unweighted case.

Main points of the section

1. It is straightforward to introduce frequency weighting into the balanced stochastic truncation algorithm. One does this by using frequency weighting when reducing the causal part of a certain all-pass function.

3.7 Sampled-data Controller Reduction

The procedures discussed to this point all assume that the plant and controller are continuous. What if the controller is discrete-time, sitting in a loop like that shown in Figure 3.7.1?

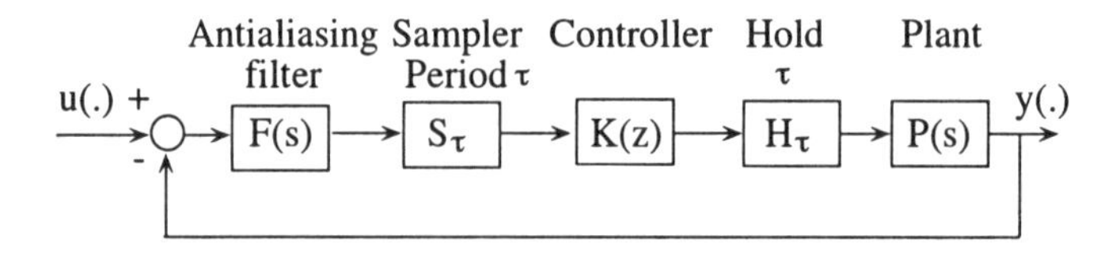

Figure 3.7.1. Sampled-data closed loop

It is well known that there is a discrete time representation of the plant which captures the behaviour of the sampled continuous time plant, see Figure 3.7.2.

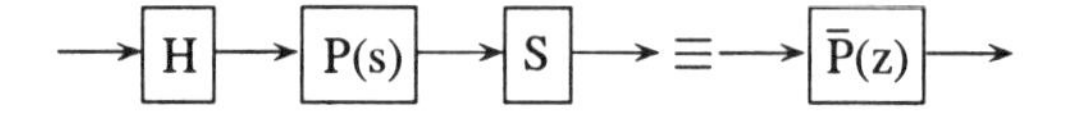

Figure 3.7.2. Replacing the continuous-time plant by a discrete-time plant.

Such a replacement may allow us to reduce the order of a high order $K(z)$ by using a surrogate closed-loop as in Figure 3.7.3. Simple variants of the earlier ideas of the chapter will handle the discrete-time nature of the problem.

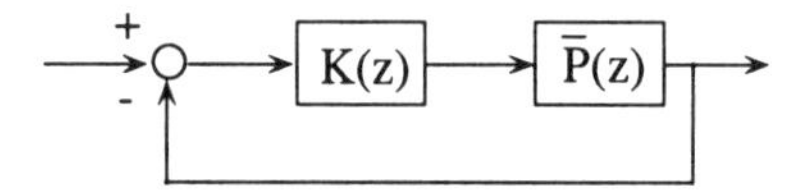

Figure 3.7.3. Discrete-time closed-loop system

However, all intersample behaviour at the plant output is lost and it may be that a reduction procedure is needed which better captures the closed-loop behaviour of Figure 3.7.1. Such a scheme is to be found in Madievski and Anderson (1995).

The core idea is to introduce fast sampling to the loop of Figure 3.7.1 (as a multiple of the underlying sampling frequency). This introduces a discrete-time periodically time-varying system. Then a "lifting" procedure produces a discrete-time time-invariant system. The controller $K(z)$ is then reduced, attempting to preserve the closed-loop behaviour of this system.

Fast sampling and lifting

Let N be an integer. Consider the arrangement of Figure 3.7.4.

It is intuitively clear that the scheme will behave in a like manner to that of Figure 3.7.1—the more so if N is large. (This statement can be made precise—it is really a statement about an approximation error tending to zero as $N \to \infty$ but we shall not pursue this here.) We shall aim to approximate $K(z)$ to preserve the behaviour from u_d to y_d in Figure 3.7.4

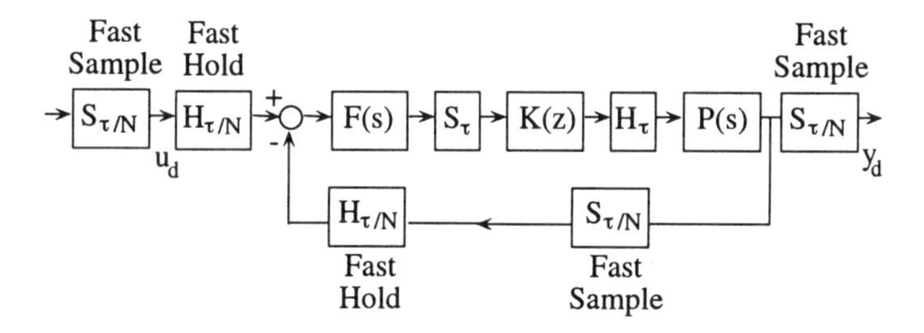

Figure 3.7.4. Fast-sampled closed-loop system

Suppose that $u_d \in R^m$, $y_d \in R^p$. To eliminate the periodic time-varying nature of Figure 3.7.4, we engage in a process of blocking. This involves assembling of N sequential values of u_d and of y_d into vectors $U_d \in R^{mN}$ and $Y_d \in R^{pN}$ and regarding the set up of Figure 3.7.4 as transforming a U_d sequence into a Y_d sequence. One new U_d and Y_d pair arises each τ seconds. Thus

$$U_d(k\tau) = \begin{bmatrix} u_d(k\tau) \\ u_d(k\tau + \frac{1}{N}\tau) \\ \vdots \\ u_d(k\tau + \frac{(N-1)\tau}{N} \end{bmatrix}, \qquad Y_d(k\tau) = \begin{bmatrix} y_d(k\tau) \\ y_d(k\tau + \frac{1}{N}\tau) \\ \vdots \\ y_d(k\tau + \frac{(N-1)\tau}{N}) \end{bmatrix}. \tag{3.7.1}$$

The block diagram representation of this is shown in Figure 3.7.5.

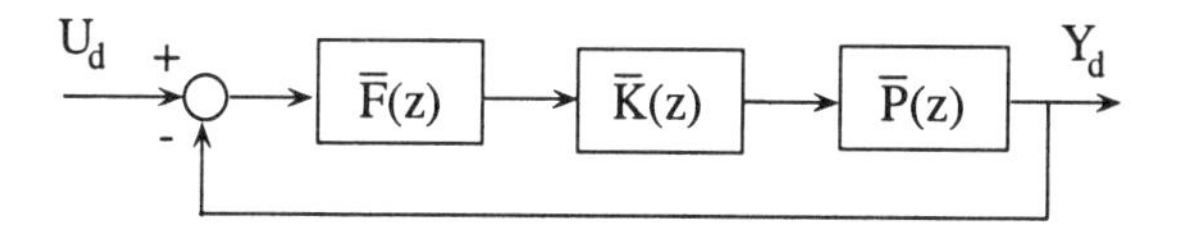

Figure 3.7.5. Representation of previous equation

In more detail, suppose the plant and antialiasing filter are

$$P(s) = C_p(sI - A_p)^{-1}B_p + D_p, \tag{3.7.2}$$

$$F(s) = C_f(sI - A_f)^{-1}B_f. \tag{3.7.3}$$

Then

$$\bar{P}(z) = \bar{C}_p(zI - \bar{A}_p)^{-1}\bar{B}_p + \bar{D}_p, \tag{3.7.4}$$

$$\bar{F}(z) = \bar{C}_f(zI - \bar{A}_f)^{-1}\bar{B}_f + \bar{D}_f, \tag{3.7.5}$$

where

$$\bar{A}_p = a_p^N, \qquad \bar{B}_p = \begin{bmatrix} a_p^{N-1} b_p & \dots & a_p b_p & b_p \end{bmatrix}, \tag{3.7.6}$$

$$\bar{A}_f = a_f^N, \qquad \bar{B}_f = \begin{bmatrix} a_f^{N-1} b_f & \dots & a_f b_f & b_f \end{bmatrix}, \tag{3.7.7}$$

$$\bar{C}_p = \begin{bmatrix} C_p^T & a_p^T C_p^T & \dots & (a_p^T)^{N-1} C_p^T \end{bmatrix}^T, \tag{3.7.8}$$

$$\bar{C}_f = \begin{bmatrix} C_f^T & a_f^T C_f^T & \dots & (a_f^T)^{N-1} C_f^T \end{bmatrix}^T, \tag{3.7.9}$$

$$\bar{D}_p = \begin{bmatrix} D_p & 0 & \dots & 0 \\ C_p b_p & D_p & \ddots & \vdots \\ \vdots & & \ddots & 0 \\ C_p a_p^{N-2} b_p & C_p a_p^{N-3} b_p & \dots & D_p \end{bmatrix}, \tag{3.7.10}$$

$$\bar{D}_f = \begin{bmatrix} 0 & 0 & \dots & 0 \\ C_f b_f & 0 & \dots & 0 \\ \vdots & & \ddots & \vdots \\ C_f a_f^{N-2} b_f & C_f a_f^{N-3} b_f & & 0 \end{bmatrix}, \tag{3.7.11}$$

$$a_p = \exp(A_p \tau / N), \qquad a_f = \exp(A_f \tau / N), \tag{3.7.12}$$

$$b_p = \int_0^{\tau/N} \exp(A_p t) dt B_p, \qquad b_f = \int_0^{\tau/N} \exp(A_f t) dt B_f. \tag{3.7.13}$$

Finally

$$\bar{K}(z) = E_1 K(z) E_2. \tag{3.7.14}$$

where

$$E_1 = \begin{bmatrix} I_m & I_m & \dots & I_m \end{bmatrix}^T \in R^{mN \times m}, \qquad E_2 = \begin{bmatrix} I_p & 0_p & \dots & 0_p \end{bmatrix} \in R^{p \times pN}. \tag{3.7.15}$$

Effectively E_2 corresponds to the slow sampler S_τ at the input of $K(z)$. This sampler only picks up one out of each N fast samples spaced τ/N seconds apart, at the fast sampler output of $F(s)$. Also, E_1 corresponds to the hold at the output of $K(z)$. Within an interval of τ seconds, there are N samples τ/N seconds apart at the hold output which will be identical.

Now define

$$\tilde{P} = \bar{P} E_1, \qquad \tilde{F} = E_2 \bar{F}. \tag{3.7.16}$$

We have for the closed-loop of Figure 3.7.4 a closed-loop transfer function matrix

$$\tilde{T} = \tilde{P} K \tilde{F} \big(I + \tilde{P} K \tilde{F}\big)^{-1}. \tag{3.7.17}$$

Controller reduction

While $\tilde{T}$ is not exactly of the same form as closed-loop transfer functions involving K which we have earlier encountered, there is an obvious way to proceed. Let

$$\hat{T} = \tilde{P}\hat{K}\tilde{F}\big(I + \tilde{P}\hat{K}\tilde{F}\big)^{-1}. \tag{3.7.18}$$

Then

$$\tilde{T} - \hat{T} \simeq \big(I + \tilde{P}K\tilde{F}\big)^{-1}\tilde{P}\big(K - \hat{K}\big)\tilde{F}\big(I + \tilde{P}K\tilde{F}\big)^{-1}, \tag{3.7.19}$$

which suggests the choice of weighting functions

$$W = \big(I + \tilde{P}K\tilde{F}\big)^{-1}\tilde{P}, \qquad V = \tilde{F}\big(I + \tilde{P}K\tilde{F}\big)^{-1} \tag{3.7.20}$$

and the index measuring approximation error:

$$J = \left\| W\big(K - \hat{K}\big)V \right\|_{\infty}. \tag{3.7.21}$$

The task of finding a reduced order $\hat{K}$ can be treated by frequency-weighted balanced truncation, with the unstable part of K being copied into $\hat{K}$.

The reference Madievski and Anderson (1995) considers sampled-data control of a system containing four spinning disks, connected by a flexible rod. A motor applies torque to the third disk, and the angular displacement of the first disk is measured. An eighth order controller is reduced to order 2. When $N = 1$ in the procedure above, no fast sampling is employed. The reference compares $N = 1$ and $N = 3$ and $N = 10$. The latter two give almost identical closed-loop performance. The performance is then far closer to the performance with the full order controller than is the performance obtained with the reduced order controller derived using $N = 1$.

Main points of the section

1. Sampled-data controller reduction seeks discrete-time controller simplification while preserving continuous-time closed-loop performance.

2. Fast sampling of the continuous-time loop together with a lifting procedure allows the sampled-data loop to be approximated by a discrete-time loop with large dimension inputs and outputs.

3. Controller reduction can occur, aimed at preserving the discrete-time loop performance.

4. More accurate reductions are to be expected than if the plant is simply replaced by its discrete-time equivalent.

3.8 Example

Let us compare some closed loop designs which consist of a plant and various reduced order controllers. The physical plant comprises four spinning disks, connected by a flexible rod, with torque applied to the third disk. Angular displacement of the first disk is of interest. The plant is modelled by an eighth order system which has a double integrator and three vibration modes. A minimal realization of the plant in modal coordinates is given by

$$A = \text{diag}\left\{\begin{bmatrix} 0 & 1 \\ 0 & 0 \end{bmatrix}, \begin{bmatrix} -0.015 & 0.765 \\ -0.765 & -0.015 \end{bmatrix}, \begin{bmatrix} -0.028 & 1.410 \\ -1.410 & -0.028 \end{bmatrix}, \begin{bmatrix} -0.04 & 1.85 \\ -1.85 & -0.04 \end{bmatrix}\right\},$$

$$B = \begin{bmatrix} 0.026 & -0.251 & 0.033 & -0.886 & -4.017 \times 2 & 0.145 \times 2 & 3.604 & 0.280 \end{bmatrix},$$

$$C = \begin{bmatrix} -0.996 \times 3 & -0.105 \times 3 & 0.261 & 0.009 & -0.001 & -0.043 & 0.002 & -0.026 \end{bmatrix}.$$

Controller reduction for a similar type of plant has been studied in Enns (1984), Anderson and Moore (1989) and Zhou *et al.* (1996). The poles and zeros are given in Table 3.8.1. (Note the existence of nonminimum phase zeros.)

Table 3.8.1. Disk system poles and zeros

Poles	Zeros
$p_1, p_2 = 0$	$z_1, z_2 = -0.017\,150\,64 \pm j0.845\,423\,8$
$p_3, p_4 = -0.015 \pm j0.765$	$z_3, z_4 = -0.539\,863\,7 \pm j1.922\,102$
$p_5, p_6 = -0.028 \pm j1.41$	$z_5, z_6 = 0.412\,711\,8 \pm j1.965\,312$
$p_7, p_8 = -0.04 \pm j1.85$	$z_7 = 104.397\,6$

Let $K(s)$ be the controller and consider a loop shaping design on $G(s)K(s)$ to secure the closed loop performance and robustness against nonstructured uncertainty. The design specification is given by the constraints on the loop gain $|G(s)K(s)|$, depicted with a hatched area in Figure 3.8.1. We obtain a high order LQG controller $K(s)$ by tuning the weightings in the quadratic performance index J_{LQ} and the covariance matrix of the noise disturbance. The designed loop gain meets the constraints, as shown in Figure 3.8.1.

Let us compare four methods for obtaining fourth order controllers in this example. The first method is the simplest one; the plant $G(s)$ is reduced to a fourth order model $\hat{G}_{mt}(s)$ by mode truncation. We shall simply neglect the truncated variables in the quadratic performance index J_{LQ} to make the new index $\hat{J}_{LQ}$ for the reduced order plant $\hat{G}_{mt}(s)$. The reduced order LQG controller $\hat{K}_{mt}(s)$ is designed with the data $(\hat{G}_{mt}(s), \hat{J}_{LQ})$. We can confirm that the constraints on the loop gain are satisfied with the designed fourth order controller $\hat{K}_{mt}(s)$ and the fourth order plant model $\hat{G}_{mt}(s)$. (See Figure 3.8.1.)

The second method uses multiplicative approximation with frequency weighting as described in Section 3.6. We have first to select the frequency weighting $V(s)$ in this method; $V(s)$ should have low pass characteristics, and the bandwidth is selected to be about the cut off frequency of the closed loop system. Accordingly we set

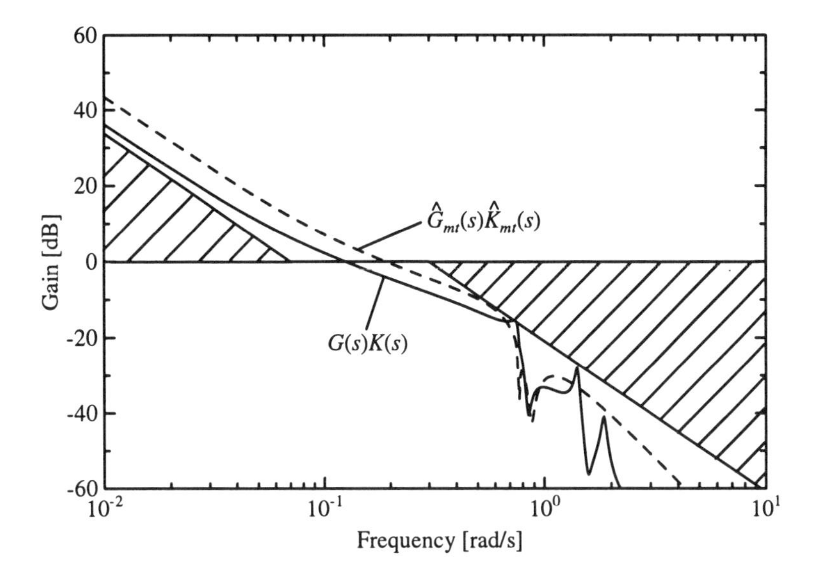

Figure 3.8.1. Loop gain design constraint with results of two LQG designs

$V(s) = 1/(s+0.25)$ based on the constraints on the loop gain. Then, we seek to obtain the fourth order plant $\hat{G}_{mp}(s)$ which minimizes $J = \|G^{-1}(G - \hat{G}_{mp})V\|_\infty$. This is approximately done through balanced stochastic truncation and bilinear transformations for treating the singularities of $G(\infty)$ and $V(\infty)$, as is explained in Section 3.6. In the balanced stochastic truncation procedure, we use a nonsingular transformation $x = Tz$ which corresponds to producing a balanced realization of $\{A, B, K\}$. (This is the realization of the stable part of an all-pass, whose additive reduction induces the multiplicative reduction we seek, see Chapter 2.) The state variables of the truncated model correspond to z_1 where $\left(z = [z_1^T\ z_2^T]^T\right)$. If the truncated model approximates the original system well, this means that the z_2 part of the original system can be neglected. This reasoning leads to the performance index of the reduced order plant $\hat{G}_{mp}(s)$ as follows:

$$\hat{J}_{LQ} = \int_0^\infty \left(\hat{x}^T \left[T^T Q T\right]_{11} \hat{x} + u^T R u\right) dt$$

where $[T^T Q T]_{11}$ indicates the (1.1) block matrix of $T^T Q T$. The designed fourth order controller $\hat{K}_{mp}(s)$ also meets the constraints with $\hat{G}_{mp}(s)$. (This is not depicted.) The original plant $G(s)$ and the reduced order plants $\hat{G}_{mt}(s)$ and $\hat{G}_{mp}(s)$ are compared in gain characteristics, in Figure 3.8.2. We can see the two methods try to copy the first gain peak of the original plant.

The third and fourth methods are squarely based on closed loop considerations. Consider the frequency weighted index $J = \|(K - \hat{K})G(I + GK)^{-1}\|_\infty$ where K is the high order controller designed for the original plant G. We can carry out an approximate minimization of J using frequency weighted balanced truncation and

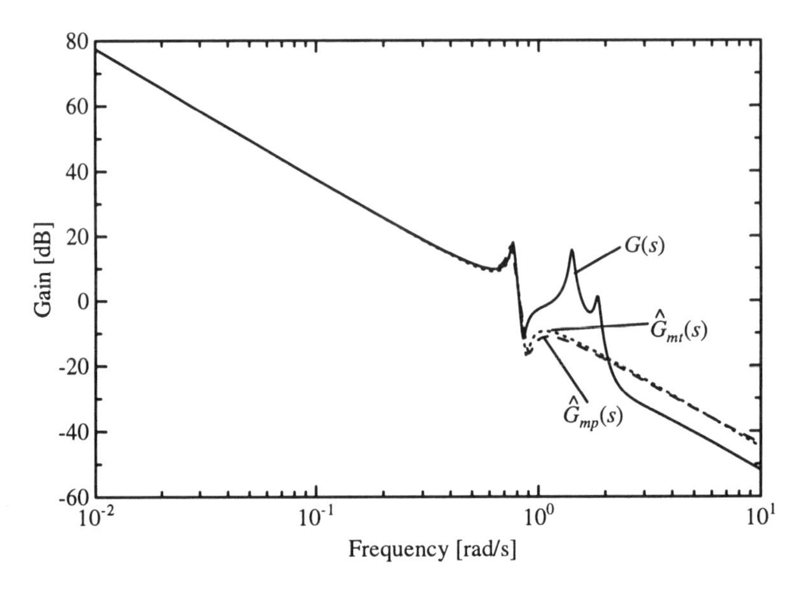

Figure 3.8.2. Gain characteristics of original plant and two reduced order models

Hankel norm approximation. In the case of weighted Hankel norm approximation, we need to modify the weighting $V = G(I + GK)^{-1}$ to have nonsingular $V(\infty)$. Let us denote the designed controllers $\hat{K}_{wbt}(s)$, $\hat{K}_{woh}(s)$, respectively. Figure 3.8.3 shows the comparison of the four methods in the loop gains (computed with the full order plant). We can confirm that the results with $\hat{K}_{wbt}(s)$ and $\hat{K}_{woh}(s)$ match well to the high order design in the low frequency region and violate the constraints a little around 1.4 rad/s, while the results with $\hat{K}_{mt}(s)$ and $\hat{K}_{mp}(s)$ fail to match in the low frequency region and break the constraints significantly around 1.4 rad/s. Figure 3.8.4 shows the gain plots of the designed controllers. The gains of $\hat{K}_{wbt}(s)$ and $\hat{K}_{woh}(s)$ match well to that of $K(s)$ in the low frequency region and are lower except around 1.6 rad/s while the gains of $\hat{K}_{mt}(s)$ and $\hat{K}_{mp}(s)$ take higher values than that of $K(s)$ except around 0.77 rad/s. Figure 3.8.5 shows the gain plots of the closed loop transfer functions $GK(I+GK)^{-1}$ and $G\hat{K}(I + G\hat{K})^{-1}$. $\hat{K}_{wbt}(s)$ and $\hat{K}_{woh}(s)$ retain a low gain in the high frequency region similarly to the high order design; therefore, those controllers secure robustness against unstructured uncertainty. The values of J are 1.366 69, 1.042 888, 0.114 535 and 0.148 794 for $\hat{K}_{mt}(s)$, $\hat{K}_{mp}(s)$, $\hat{K}_{wbt}(s)$ and $\hat{K}_{woh}(s)$, respectively. Though all controllers ensure the stability of the closed loop system, $\hat{K}_{wbt}(s)$ and $\hat{K}_{woh}(s)$ have a larger stability margin than the others.

One can also design a reduced order controller with the index

$$J = \left\| (I + GK)^{-1} G(K - \hat{K})(I + GK)^{-1} \right\|_{\infty}.$$

(Here, the further weighting $(I + GK)^{-1}$ is introduced into the index.) However, the results with the fourth order controller which is calculated by weighted balanced truncation are almost the same as those with the controller $\hat{K}_{wbt}$ above.

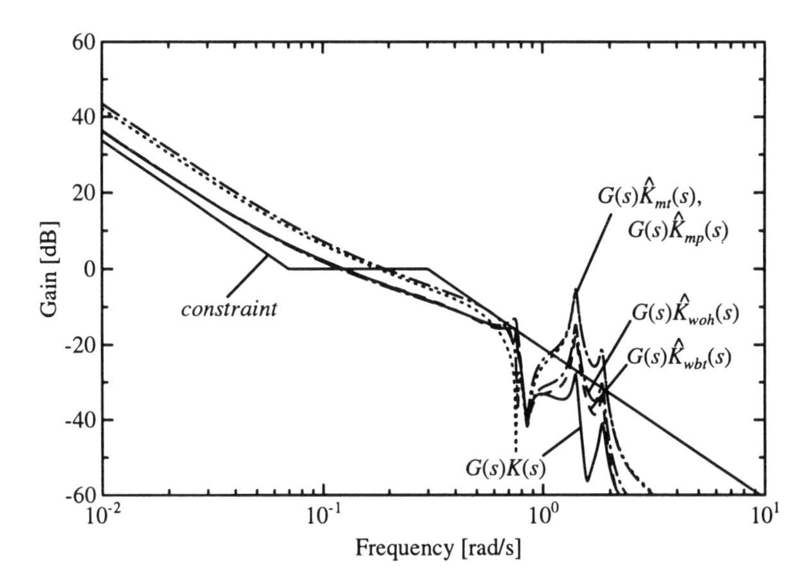

Figure 3.8.3. Comparison of loop gain with original controller and four reduced order controllers

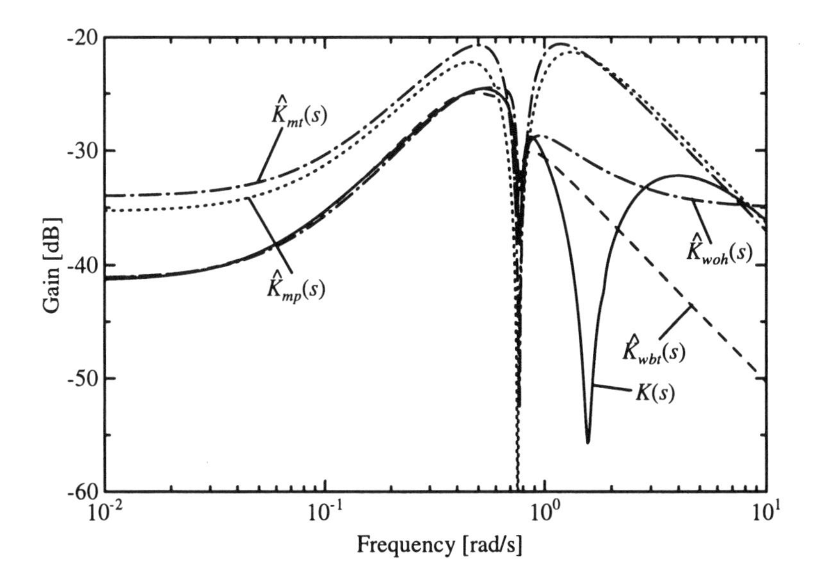

Figure 3.8.4. Gain of original controller and four reduced order controllers

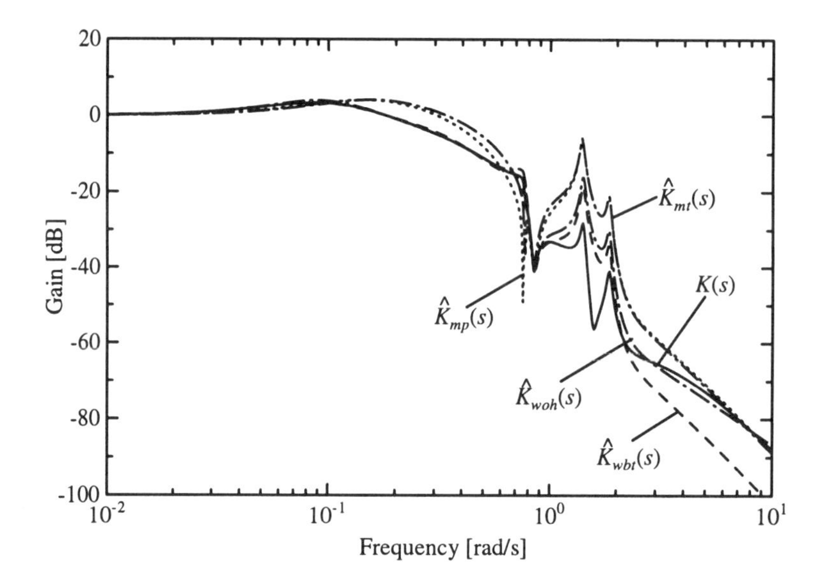

Figure 3.8.5. Closed-loop gain comparison with original controller and four reduced order controllers

This example shows that reduced order designs obtained by an indirect method starting with a high order controller are better than designs obtained by an indirect method which works through a low order plant model.

Chapter 4

Model and Controller Reduction Based On Coprime Factorizations

4.1 Coprime Factorizations: What are they and why use them?

In balanced truncation and Hankel norm reduction, any instabilities if present can only be treated by copying them into the reduced order object, which differs from the original object only through reduction of the stable part. This conclusion applies to frequency weighted reduction also, and thus to the controller reduction methods we have examined.

Similarly, in looking at multiplicative reduction, we found that in the case of the scheme tied to using Hankel norm approximation, which can cope with unstable poles in the original object, precisely these unstable poles are also present in the reduced order object.

It seems intuitively reasonable that even if high order unstable systems must be approximated by low order unstable systems, there should be no need to exactly copy the instability, and that there might in fact be some loss of optimality or quality of approximation in doing this-especially when we are talking about closed loop quantities. For this reason, we will examine methods for achieving approximations of possibly unstable transfer function matrices which do not rely on copying instabilities from the high order object to the low order object.

The tool is to use coprime fractional representations of the transfer function matrices of interest. For a full understanding of this important tool of linear system theory, the reader is encouraged to consult standard textbooks. We shall present a brief introduction in the remainder of this section and in the next section.

Consider an unstable transfer function

$$P(s) = \frac{1}{(s-1)}. \tag{4.1.1}$$

Normally we think of it as a ratio of polynomials—in this case the denominator is $s-1$ and the numerator is 1. We could however write $P(s)$ as a ratio of transfer functions each of which is proper and stable. Thus if

$$N(s) = \frac{1}{(s+1)}, \qquad D(s) = \frac{(s-1)}{(s+1)}, \tag{4.1.2}$$

it is evident that

$$P(s) = \frac{N(s)}{D(s)}, \tag{4.1.3}$$

with both $N(s)$ and $D(s)$ proper and stable. Of course, $N(s)$ and $D(s)$ are far from unique; there is no reason why the denominator of each of these stable transfer functions should not be $s+\alpha$ for some positive α, or even a higher order stable polynomial.

Actually, the possibility of using a higher order stable polynomial as the denominator of $N(s)$ and $D(s)$ is ruled out by a coprimeness condition which we shall introduce in the next section—it is analogous to ruling out descriptions of $P(s)$ as a ratio of two polynomials when the two polynomials are permitted to have a common factor, *i.e.*, it is analogous to insisting that the two polynomials in the ratio be coprime.

On the other hand, if we focus on common denominators for $N(s)$ and $D(s)$ of the form $s+\alpha$, the freedom in α is basic. There is however a "preferred" value of α, *viz.* that which ensures the following normalization condition is satisfied:

$$|N(j\omega)|^2 + |D(j\omega)|^2 = 1.$$

This is achieved by a unique positive value of α, *viz.* $\alpha = \sqrt{2}$. The pair $N(s)$, $D(s)$ with this α is termed a normalized coprime realization; the pair is unique up to multiplication by -1.

The reader familiar with the concepts of coprime fractional descriptions can omit the next section.

Main points of the section

1. Coprime fractional descriptions can be applied on reduction problems, and can be used to achieve approximations of possibly unstable transfer functions without copying instabilities from the higher order object to the lower order object. Such descriptions represent transfer functions as ratios, not of polynomials, but of proper and stable transfer functions.

4.2 Coprime Fractional Descriptions

The following material is standard, and summarizes essential facts regarding coprime fractional descriptions.

Let $P(s)$ be a real rational transfer function matrix. Any writing of $P(s)$ in the form

$$P(s) = M(s)N^{-1}(s) \tag{4.2.1}$$

or

$$P(s) = \bar{N}^{-1}(s)\bar{M}(s), \tag{4.2.2}$$

where M, N, $\bar{M}$ and $\bar{N}$ lie in RH_∞ is termed a fractional description of $P(s)$.

In fact, (4.2.1) is termed a right fractional description and (4.2.2) a left fractional description. In the material following, we will sometimes focus just on right fractional descriptions since the ideas for left fractional descriptions can usually be obtained by transposition.

A right fractional description is termed coprime if there exist $\bar{X}(s), \bar{Y}(s) \in RH_\infty$ such that a so-called Bezout identity holds:

$$\bar{X}(s)M(s) + \bar{Y}(s)N(s) = I \tag{4.2.3}$$

or equivalently if there exist $\bar{X}(s), \bar{Y}(s) \in RH_\infty$ such that

$$\bar{X}(s)M(s) + \bar{Y}(s)N(s) = Z(s), \tag{4.2.4}$$

where $Z(s)$ is a unit, *i.e.*, $Z(s) \epsilon RH_\infty$, $Z^{-1}(s) \epsilon RH_\infty$. [Note that (4.2.4) holding implies (4.2.3) holds with $\bar{X}$, $\bar{Y}$ replaced by $Z^{-1}\bar{X}$, $Z^{-1}\bar{Y}$, which are again in RH_∞.]

An alternative and equivalent statement of coprimeness is that

$$R(s) = \begin{bmatrix} M(s) \\ N(s) \end{bmatrix} \tag{4.2.5}$$

has full rank at every point in the closed right half plane. If MN^{-1} is a right coprime realization of $P(s)$, all others are given by $(MZ)(NZ)^{-1}$ where Z is a unit but is otherwise arbitrary.

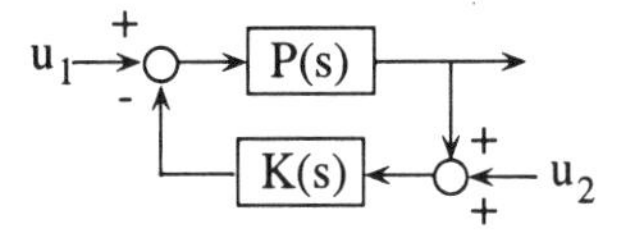

Figure 4.2.1. Closed-loop negative feedback system

Let $P(s) = M(s)N^{-1}(s)$ and $K(s) = \bar{Y}^{-1}(s)\bar{X}(s) = X(s)Y^{-1}(s)$ provide a right coprime fractional description of a plant and left and right coprime fractional descriptions of a controller forming an associated closed loop negative feedback system, see Figure 4.2.1. Then this system is closed-loop stable if and only if (4.2.4) holds for

$Z(s)$ a unit, and also if and only if

$$\Psi(s) = \begin{bmatrix} Y(s) & M(s) \\ -X(s) & N(s) \end{bmatrix} \tag{4.2.6}$$

is a unit. More generally, let

$$P = MN^{-1} = \bar{N}^{-1}\bar{M}, \qquad K = \bar{Y}^{-1}\bar{X} = XY^{-1}, \tag{4.2.7}$$

define arbitrary coprime realizations. Then, if and only if there is closed-loop stability,

$$\begin{bmatrix} \bar{N} & -\bar{M} \\ \bar{X} & \bar{Y} \end{bmatrix} \begin{bmatrix} Y & M \\ -X & N \end{bmatrix} = \begin{bmatrix} \bar{Z}(s) & 0 \\ 0 & Z(s) \end{bmatrix}, \tag{4.2.8}$$

where Z and $\bar{Z}$ are units. By replacing the coprime description MN^{-1} by the coprime description $(MZ)^{-1}(NZ^{-1})^{-1}$ (remember that Z and so Z^{-1} are units), and by replacing the coprime description XY^{-1} in a similar vein by $(X\bar{Z}^{-1})(Y\bar{Z}^{-1})^{-1}$, we see that we can use coprime realizations such that

$$\begin{bmatrix} \bar{N} & -\bar{M} \\ \bar{X} & \bar{Y} \end{bmatrix} \begin{bmatrix} Y & M \\ -X & N \end{bmatrix} = I. \tag{4.2.9}$$

This is termed the double Bezout identity. We are saying that if the closed loop of Figure 4.2.1 is stable *special* coprime realizations of $P(s)$ and $K(s)$ can be found so that (4.2.9) holds; and if for certain coprime realizations of $P(s)$ and $K(s)$ equation (4.2.9) holds, then the closed loop is stable.

As already noted, if MN^{-1} is a right coprime realization of $P(s)$, so is $(MZ) \times (NZ)^{-1}$ where Z is a unit but is otherwise arbitrary. Choose $Z(s)$ to satisfy

$$M_*(s)M(s) + N_*(s)N(s) = Z_*(s)Z(s). \tag{4.2.10}$$

This means that $Z(s)$ is a stable, minimum phase spectral factor of the power spectrum on the left side of (4.2.10); this power spectrum is nonsingular on the imaginary axis when MN^{-1} is coprime—observe the left side of (4.2.10) is $R^T(-s)R(s)$ with $R(s)$ in (4.2.5) of full rank on the imaginary axis.

With this choice of $Z(s)$, the new fractional description, call it $\hat{M}(s)\hat{N}^{-1}(s)$ say, satisfies

$$\hat{M}_*(s)\hat{M}(s) + \hat{N}_*(s)\hat{N}(s) = I. \tag{4.2.11}$$

Such a fractional description is termed normalized. Normalized right coprime fractional descriptions are unique to within right multiplication of their constituents by a constant orthogonal matrix.

In a similar manner normalized left coprime fractional descriptions can be found, and they are unique to within left multiplication of their constituents by a constant orthogonal matrix.

Above, we have pointed out that one or both of the left and right coprime fractional description of a plant and a controller can be normalized (but do not have to be), and that the four coprime fractional description associated with a plant and stabilizing controller may satisfy the double Bezout identity (4.2.9) (but do not have to, though (4.2.8) must be satisfied for units Z and $\bar{Z}$). The question arises: how many of the normalization and Bezout requirements can be simultaneously imposed? The answer is that any of the following possibilities can be achieved, but in general no more.

1. All four fractions are normalized
2. The left fractions are both normalized (or the right fractions are both normalized) and the double Bezout identity holds
3. The fractional descriptions of the plant are both normalized (or the fractional description of the controller are both normalized) and the double Bezout identity holds

State-variable constructions of the Bezout identity

Many of the above ideas have nice state-space interpretations; we shall now present some of these. Let

$$P(s) = C(sI - A)^{-1}B. \tag{4.2.12}$$

Let F be a matrix such that the eigenvalues of $A + BF$ lie in the open left half plane. If (A, B) is stabilizable, such an F always exists. Then

$$P(s) = C(sI - A - BF)^{-1}B\big[I + F(sI - A - BF)^{-1}B\big]^{-1}. \tag{4.2.13}$$

So we can take

$$\begin{bmatrix} M(s) \\ N(s) \end{bmatrix} = \left[\begin{array}{c|c} A + BF & B \\ \hline C & 0 \\ F & I \end{array}\right]. \tag{4.2.14}$$

Is this fractional description coprime? Some elementary system theory shows that it is coprime if and only if the pair (A, C) is detectable. [In brief, suppose for some s_0 with $\mathrm{Re}[s_0] \geq 0$ and some complex vector $\alpha \neq 0$, there holds $M(s_0)\alpha = 0$, $N(s_0)\alpha = 0$. Let $\beta = [s_0 I - (A + BF)]^{-1}B\alpha$. The inverse necessarily exists because $\mathrm{Re}[s_0] \geq 0$ and $A+BF$ has open left half plane eigenvalues. Now $N(s_0)\alpha = 0$ implies $F\beta + \alpha = 0$ and so $F\beta \neq 0$. Hence $\beta \neq 0$. Also, the definition of β yields $(s_0 I - A - BF)\beta - B\alpha = 0$, whence $(s_0 I - A)\beta = 0$. The requirement that $M(s_0)\alpha = 0$ also yields $C\beta = 0$. This violates detectability. Evidently then, if $\{A, B, C\}$ in (4.2.12) is a minimal triple, the above procedure will lead to a coprime fractional description.

Next, let $P(s)$ be as in (4.2.12) with $\{A, B, C\}$ stabilizable and detectable. Let F and H be matrices such that $A + BF$ and $A + HC$ have all eigenvalues in the open left half plane. Using F, a stabilizing state feedback controller can be obtained for

$$\dot{x} = Ax + Bu, \qquad y = Cx \tag{4.2.15}$$

and using H, a state estimator can be obtained:

$$\dot{\hat{x}} = (A + HC)\hat{x} + Bu - Hy. \tag{4.2.16}$$

Using F and H together, a stabilizing output feedback controller with transfer function matrix $K(s)$ can be obtained, based on feeding back $F\hat{x}$:

$$\dot{\hat{x}} = (A + HC + BF)\hat{x} - Hy, \qquad u = F\hat{x}. \tag{4.2.17}$$

Allowing for a negative feedback connection, it is evident that the transfer function $K(s)$ is given by $F(sI - A - HC - BF)^{-1}H$. (The language we are using suggests that $P(s)$ is a plant; it does not have to be. The words "state feedback controller" and the like are used to motivate the construction of $K(s)$ and certain associated state-variable descriptions.)

Now set

$$M(s) = C(sI - A - BF)^{-1}B, \quad N(s) = I + F(sI - A - BF)^{-1}B, \tag{4.2.18}$$
$$\bar{M}(s) = C(sI - A - HC)^{-1}B, \quad \bar{N}(s) = I + C(sI - A - HC)^{-1}H, \tag{4.2.19}$$

so that

$$P(s) = M(s)N^{-1}(s) = \bar{N}^{-1}(s)\bar{M}(s). \tag{4.2.20}$$

Also set

$$X(s) = F(sI - A - BF)^{-1}H, \quad Y(s) = I - C(sI - A - BF)^{-1}H, \tag{4.2.21}$$
$$\bar{X}(s) = F(sI - A - HC)^{-1}H, \quad \bar{Y}(s) = I - F(sI - A - HC)^{-1}B. \tag{4.2.22}$$

Then

$$K(s) = F(sI - A - HC - BF)^{-1}H = X(s)Y^{-1}(s) = \bar{Y}^{-1}(s)\bar{X}(s). \tag{4.2.23}$$

Furthermore, the double Bezout identity (4.2.9) holds with these choices:

$$\begin{bmatrix} I + C(sI - A - HC)^{-1}H & -C(sI - A - HC)^{-1}B \\ F(sI - A - HC)^{-1}H & I - F(sI - A - HC)^{-1}B \end{bmatrix} \times \begin{bmatrix} I - C(sI - A - BF)^{-1}H & C(sI - A - BF)^{-1}B \\ -F(sI - A - BF)^{-1}H & I + F(sI - A - BF)^{-1}B \end{bmatrix} = I. \tag{4.2.24}$$

For this result, see Nett, Jacobson and Balas (1984).

State-variable construction of normalized description

To conclude this section, we consider how to pass from a state variable realization $P(s) = C(sI - A)^{-1}B$ or $D + C(sI - A)^{-1}B$ to a normalized coprime fractional description, and we record an interesting property of such descriptions. First suppose that $D = 0$ and that $\{A, B, C\}$ is stabilizable and detectable. Let Q be the stabilizing solution of the Riccati equation

$$QA + A^T Q - QBB^T Q + C^T C = 0 \tag{4.2.25}$$

and select the matrix F in (4.2.13) and (4.2.14) to be $-B^T Q$. It takes a few lines of algebra to verify that

$$M^T(-s)M(s) + N^T(-s)N(s) = I. \tag{4.2.26}$$

In case $D \neq 0$, set $R = I + D^T D$, $\tilde{R} = I + DD^T$ and define Q as the solution of

$$Q\left(A - BR^{-1}D^T C\right) + \left(A - BR^{-1}D^T C\right)^T Q - QBR^{-1}B^T Q + C^T \tilde{R}^{-1} C = 0 \tag{4.2.27}$$

and

$$F = -R^{-1}\left(B^T Q + D^T C\right). \tag{4.2.28}$$

Then

$$\begin{bmatrix} M \\ N \end{bmatrix} = \left[\begin{array}{c|c} A + BF & BR^{-1/2} \\ \hline F & R^{-1/2} \\ C + DF & DR^{-1/2} \end{array}\right]. \tag{4.2.29}$$

Of course, left coprime normalized fractional descriptions can be found in the same way Evidently, the Riccati equation associated with linear quadratic control is linked with finding a right coprime normalized description, and that associated with the Kalman filter is linked with finding a left coprime normalized description.

Notice that the fractional descriptions of (4.2.20) have the same state variable dimension for the obvious realization of the transfer function matrix

$$R(s) = \begin{bmatrix} M(s) \\ N(s) \end{bmatrix}$$

as the original $P(s)$, and the normalized coprime fractional description in (4.2.29) is a special case.

Balanced truncation of normalized coprime factors

There is a simple calculation that shows that if one truncates a balanced realization of a normalized pair arranged "over" each other as in (4.2.29), the result is again normalized (Meyer, 1990). Suppose simply that

$$\begin{bmatrix} M(s) \\ N(s) \end{bmatrix} = \left[\begin{array}{c|c} A & B \\ \hline C & D \end{array}\right] \tag{4.2.30}$$

and

$$I = D^T D + D^T C(sI - A)^{-1} B + B^T(-sI - A^T)^{-1} C^T D \\ + D^T(-sI - A^T)^{-1} CC^T (sI - A)^{-1} B. \tag{4.2.31}$$

Now, noting that the observability grammian Q obeys

$$Q(sI - A) + (-sI - A^T)Q = C^T C \tag{4.2.32}$$

we can replace (4.2.31) by

$$I = D^T D + (D^T C + B^T Q)(sI - A)^{-1} B + B^T(-sI - A^T)^{-1}(C^T D + QB). \tag{4.2.33}$$

The normalization property in state variable terms is therefore

$$I = D^T D, \qquad QB + C^T D = 0. \tag{4.2.34}$$

Now if Q is a diagonal matrix Σ, the second equation in (4.2.34) yields $\Sigma_1 B_1 + C_1^T D = 0$ where Σ_1, B_1 and C_1 are obvious submatrices of Σ_1, B_1 and C_1 with the latter arising in the truncation operation. It follows that because (4.2.34) holds for the truncated reduced order system, it too obeys the normalization property.

Main points of the section

1. Coprimeness can be characterized with a Bezout identity, and also with a constancy-of-rank condition. The Bezout identity has an interpretation involving a closed-loop system with plant and feedback controller.
2. Normalized coprime fractional descriptions can be constructed, and are unique up to multiplication of their constituents by a constant orthogonal matrix.
3. Normalized coprime fractional descriptions can be constructed using state-variable descriptions and a Riccati equation solution.
4. If one truncates a balanced realization of a normalized coprime pair, the reduced order pair is again normalized.

4.3 Reducing Controller Dimension Using Coprime Fractions

Let $K(s)$ be the transfer function matrix of a high order controller. Suppose that $X(s)Y^{-1}(s)$ is a coprime fractional description of the controller. Set

$$R(s) = \begin{bmatrix} X(s) \\ Y(s) \end{bmatrix}.$$

Can we find a reduced order controller by reducing $R(s)$ to an approximating object $\hat{R}(s)$, and then using the reduced order approximations of $X(s)$ and $Y(s)$, call them $\hat{X}(s)$ and $\hat{Y}(s)$, to form a low order approximating controller $\hat{K}(s) = \hat{X}(s)\hat{Y}^{-1}(s)$?

Obviously we can try this. But will it be profitable? Key issues include

1. What particular fractional representation is it best to use?
2. Even if we reduce the degree of $R(s)$ in forming an approximation of it $\hat{R}(s)$, can we ensure this leads to a $\hat{K}(s)$ that has lower degree than $K(s)$?
3. Should we introduce weights in the reduction process or not? If so, what should they be?

The answer to the middle question is the easiest. We shall consistently use fractional descriptions so that the matrix $R(s)$ above has identical degree with that of $K(s)$, and so that the degree of $\hat{K}(s)$ will be no greater than (and will normally be equal to) that of $\hat{R}(s)$.

We shall begin by treating controllers which have resulted from an LQG design process. It will turn out that the method developed for reducing them is applicable to controllers produced by any design technique.

Reduction using a noise-induced index

Suppose the initially given plant is

$$P(s) = C(sI - A)^{-1}B \tag{4.3.1}$$

and that an LQG design procedure is used to establish a state feedback gain matrix F and a Kalman filter gain matrix H, so that the resultant closed loop with the controller depicted in detail is as shown in Figure 4.3.1.

The controller transfer function matrix $K(s)$ has a right coprime fractional description as $X(s)Y^{-1}(s)$ where

$$X(s) = F(sI - A - BF)^{-1}H, \qquad Y(s) = I - C(sI - A - BF)^{-1}H. \tag{4.3.2}$$

Observe now that $X(s)$ and $Y(s)$ are more or less the transfer functions from v to u and v to z (assuming the feedback at z is disconnected and external signals are applied

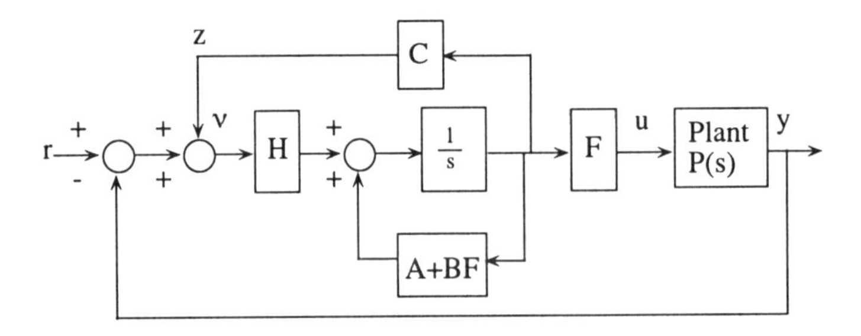

Figure 4.3.1. Controller-plant loop resulting from LQG design

at ν) Recall also that in an ideal LQG design, the signal ν (known as the innovations) is white, *i.e.*, it has constant spectrum, see (Anderson and Moore, 1989). If this is so, it provides a heuristic basis for approximating $X(s)$ and $Y(s)$ *with no weighting*—all frequencies are equally important as each other, when it comes to looking at the input of $X(s)$ and $Y(s)$.

Another way of considering the approximation issue is to say that temporarily we remove the feedback connection at z, we approximate the controller minus the feedback connection (and this involves approximating $X(s)$ and $Y(s)$ by some $\hat{X}(s)$ and $\hat{Y}(s)$), and then we restore the feedback connection with the approximating entities in place.

Above, we have presented a heuristic argument justifying the use of no weighting in the reduction procedure, when the controller has been formed using LQG design ideas. We shall now indicate an alternative justification, based on certain non-LQG performance considerations, due to Zhou and Chen (1995).

Consider the scheme of Figure 4.3.2, showing as it does two separate input and output points.

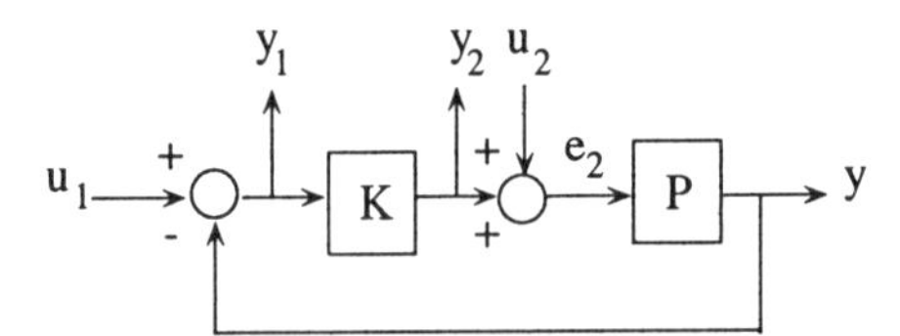

Figure 4.3.2. MIMO Feedback System

Consider also the transfer function matrix

$$T(P,K) = \begin{bmatrix} I \\ K \end{bmatrix} (I + PK)^{-1} \begin{bmatrix} I & P \end{bmatrix}. \tag{4.3.3}$$

It is easy to check that

$$\begin{pmatrix} y_1 \\ y_2 \end{pmatrix} = T(P,K) \begin{bmatrix} u_1 \\ -u_2 \end{bmatrix} \tag{4.3.4}$$

and $T(P, K)$ carries within it four transfer function matrices that determine the closed-loop behaviour. (Note incidentally that the transfer function from u_1 to y is given by $1 - [T(P, K)]_{11}$).

It is also possible to show, see Georgiou and Smith (1990), that

$$\left\|T(P, K)\right\|_{\infty} = \left\|I - T(P, K)\right\|_{\infty}. \tag{4.3.5}$$

We shall use this property below.

Now if we have coprime fractional descriptions $P = \bar{N}^{-1}\bar{M}$ and $K = XY^{-1}$, and if we set $\bar{Z} = \bar{M}X + \bar{N}Y$, it is easily seen that

$$T_0(P, K) = \begin{bmatrix} Y \\ X \end{bmatrix} \bar{Z}^{-1} \begin{bmatrix} \bar{N} & \bar{M} \end{bmatrix}. \tag{4.3.6}$$

Suppose now that $\hat{K} = \hat{X}\hat{Y}^{-1}$ is a reduced order controller for P, and

$$\Delta_x = \hat{X} - X, \qquad \Delta_y = \hat{Y} - Y. \tag{4.3.7}$$

Then some lines of algebra yield the following:

$$T(P, \hat{K}) \tag{4.3.8}$$
$$= T(P, K) + (I - T(P, K)) \begin{bmatrix} \Delta_y \\ \Delta_x \end{bmatrix} \bar{Z}^{-1} \begin{bmatrix} \bar{N} & \bar{M} \end{bmatrix} \left[I + \begin{bmatrix} \Delta_y \\ \Delta_x \end{bmatrix} \bar{Z}^{-1} \begin{bmatrix} \bar{N} & \bar{M} \end{bmatrix} \right]^{-1}.$$

Now let us constrain the coprime fractional descriptions in two ways. First, $[\bar{N} \; \bar{M}]$ should be normalized. Second, $\bar{Z}$ should be the identity. This is indeed possible as we now argue. Recall the LQG design which is supposed to have led to the controller. With $P(s) = C(sI - A)^{-1}B$ and with S the steady state solution of the Kalman filter Riccati equation

$$SA^T + SA - SC^TCS + BB^T = 0, \tag{4.3.9}$$

the Kalman filter gain is $H = -SC^T$, and it results in $A + HC$ having eigenvalues in the open left half plane. Moreover, with

$$\bar{M}(s) = C(sI - A - HC)^{-1}B, \qquad \bar{N}(s) = I + C(sI - A - HC)^{-1}H, \tag{4.3.10}$$

one can deduce from the Riccati equation the normalization condition

$$\bar{M}(s)\bar{M}_*(s) + \bar{N}(s)\bar{N}_*(s) = I. \tag{4.3.11}$$

If F is a stabilizing state feedback gain, not necessarily but possibly obtained from a linear quadratic law, then the choices

$$X(s) = F(sI - A - BF)^{-1}H, \qquad Y(s) = I - C(sI - A - BF)^{-1}H, \tag{4.3.12}$$

yield $K(s) = X(s)Y^{-1}(s)$, and by virtue of the double Bezout identity of Section 4.2 (see (4.2.24) in particular), also $\bar{Z}(s) = \bar{M}X + \bar{N}Y = I$.

Let us also suppose that

$$\left\| \begin{bmatrix} \Delta_y \\ \Delta_x \end{bmatrix} \right\|_\infty \leq \epsilon < 1. \tag{4.3.13}$$

(When $\epsilon < 1$, it is possible to show that the reduced order controller necessarily stabilizes the plant). Then (4.3.7) yields

$$\left\| T(P, \hat{K}) - T(P, K) \right\|_\infty \leq \left\| I - T(P, K) \right\|_\infty \epsilon \left\| \bar{N} \quad \bar{M} \right\|_\infty \frac{1}{1 - \epsilon \left\| \bar{N} \quad \bar{M} \right\|_\infty}. \tag{4.3.14}$$

The normalization condition for $[\bar{N} \ \bar{M}]$ ensures that $\|\bar{N} \ \bar{M}\|_\infty = 1$, and so, using also (4.3.5), we have

$$\left\| T(P, \hat{K}) - T(P, K) \right\|_\infty \leq \left\| T(P, K) \right\|_\infty \frac{\epsilon}{1 - \epsilon}. \tag{4.3.15}$$

We see that the noise-induced reduction without weight of a certain fractional description of the controller in (4.3.2) is actually a reduction of a coprime fraction XY^{-1} for which the associated plant fractional description $\bar{N}^{-1}\bar{M}$ is normalized and obeys the Bezout identity with XY^{-1}. *It results in satisfaction of the closed-loop performance inequality* (4.3.15). *It is actually irrelevant as to whether the controller is designed using a linear quadratic performance index.*

Accordingly, if one were presented with simply $P(s)$ and a stabilizing $K(s)$ *designed by no matter what method*, and if one were asked to reduce $K(s)$, one could achieve this in the following way:

1. Form arbitrary coprime fraction descriptions $P(s) = N_1^{-1}(s)M_1(s)$ and $K(s) = X_1(s)Y_1^{-1}(s)$

2. Replace the description of $P(s)$ by a normalized coprime fractional description $\bar{N}^{-1}(s)\bar{M}(s)$

3. Evaluate $\bar{M}(s)X_1(s) + \bar{N}(s)Y_1(s) = \bar{Z}(s)$, and replace $X_1(s)$ and $Y_1(s)$ by $X(s) = X_1(s)\bar{Z}^{-1}(s)$ and $Y(s) = Y_1(s)\bar{Z}^{-1}(s)$.

4. Reduce $[X^T(s) \ Y^T(s)]^T$ without weighting. If the infinity norm of the reduction error, ϵ, obeys $\epsilon < 1$, the reduced order controller is stabilizing and (4.3.15) holds. Note however that if $K(s)$ is not obtained as an observer-based state feedback controller, the degree of $X(s)$ and $Y(s)$ could be high.

Main points of the section

1. Reducing the two factors of a coprime pair leads to a reduced order model of the original system.

2. The construction of an LQG controller suggests a particular pair to choose to represent the controller, and Kalman filter theory motivates a reduction without weighting.

3. An H_∞ norm bound on the additive error between closed-loop transfer functions when a controller is approximated can be expressed using a bound on the additive error between coprime pair descriptions of the two controllers.

4. There is a well motivated controller reduction procedure for nonLQG controllers based on using a fractional representation of the controller which satisfies a Bezout identity involving a normalized coprime fractional description of the plant. The error bound of Point 3 applies in a very simple way.

4.4 Controller Reduction by H_∞-balanced Truncation

In Mustafa and Glover (1991), Nagado, Shimemura, Ishida and Uchida (1991) and Nagado, Shimemura, Ishida and Uchida (1994), elegant procedures are given for reducing a controller defined by H_∞ techniques for a restricted class of H_∞ problem. One of the advantages is that an *a priori* test for the stability and performance (closed-loop gain in an H_∞ context) is available, although it can be conservative. The approaches are similar but not identical. That of Mustafa and Glover (1991), relies on identifying a reduced order plant which a reduced order controller is guaranteed to stabilize, and then ensuring that the controller also stabilizes the original plant. That of Nagado *et al.* (1991), Nagado *et al.* (1994) relies on ensuring that the reduced order controller, regarded as a variation on the full order controller, continues to stabilize the original plant.

Here, we shall focus more on the approach of Mustafa and Glover (1991). This approach incidentally is an outgrowth of that of Jonckheere and Silverman (1983), who introduced a concept of balancing of Riccati equations arising in an LQG design, and then truncating. As noted in Mustafa and Glover (1991), establishing sufficient conditions for retention of stability in LQG balancing and truncation leads to conditions which are very conservative, and the conditions are limited to scalar plants. Furthermore, it can happen in the approach of Jonckheere and Silverman (1983) that if the unreduced controller (as defined using Riccati equation solutions) is nonminimal, then the nonminimal modes are not necessarily exactly removed in finding the low order controller.

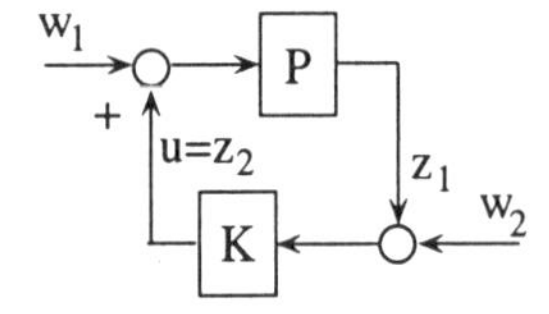

Figure 4.4.1. Closed-loop system

Consider the set-up of Figure 4.4.1. The plant is

$$\dot{x} = Ax + Bw_1 + Bu, \qquad y = Cx + w_2, \qquad z_1 = Cx, \qquad z_2 = u, \tag{4.4.1}$$

with $\{A, B, C\}$ assumed minimal, and the H_∞ -problem is to find a stabilizing feedback controller $K(s)$ linking y to u such that the gain from $[w_1^T\ w_2^T]^T$ to $[z_1^T\ z_2^T]^T$ is smaller than some prescribed value γ. The transfer function matrix is

$$\bar{T}(P, K) = \begin{bmatrix} (I - PK)^{-1}P & (I - PK)^{-1}PK \\ K(I - PK)^{-1}P & K(I - PK)^{-1} \end{bmatrix}. \tag{4.4.2}$$

It should be distinguished from the earlier introduced $T(P, K)$ of the last section, and not just in terms of the sign of the feedback for K and the ordering of inputs or outputs. In Mustafa and Glover (1991), some comment is made on this point.

In general and provided γ is large enough, the H_∞ problem does not have a unique solution but a family of solutions. However there is a so-called minimum entropy H_∞ controller which is unique. The background theory may be found in Green and Limebeer (1995) and Zhou *et al.* (1996). Let

$$\gamma_0 = \inf_{\text{stabilizing } K} \left\| \bar{T}(P, K) \right\|_\infty . \tag{4.4.3}$$

The main result is as follows.

Theorem 4.4.1. *Let $P = \{A, B, C\}$ be minimal and $\gamma > \gamma_0$. Then there exist unique positive definite symmetric stabilizing solutions X_∞ and Y_∞ to the Riccati equations*

$$A^T X_\infty + X_\infty A - \left(1 - \gamma^{-2}\right) X_\infty B B^T X_\infty + C^T C = 0, \tag{4.4.4}$$

$$A Y_\infty + Y_\infty A^T - \left(1 - \gamma^{-2}\right) Y_\infty C^T C Y_\infty + B B^T = 0 \tag{4.4.5}$$

and $\lambda_{\max}(X_\infty Y_\infty) < \gamma^2$. With

$$Z_\infty = \left(I - \gamma^{-2} X_\infty Y_\infty\right)^{-1} \tag{4.4.6}$$

the associated minimum entropy controller is

$$\begin{aligned} \dot{\hat{x}} &= A_c \hat{x} + B_c y = \left[A - \left(1 - \gamma^{-2}\right) Y_\infty C^T C - B B^T Z_\infty X_\infty\right] \hat{x} + Y_\infty C^T y, \\ u &= C_c \hat{x} = -B^T X_\infty Z_\infty \hat{x}. \end{aligned} \tag{4.4.7}$$

If one changes the coordinate basis by introducing a nonsingular transformation matrix T so that $\{A, B, C\} \to \{TAT^{-1}, TB, CT^{-1}\}$, then $X_\infty \to T^{-T}X_\infty T^{-1}$ and $Y_\infty \to TY_\infty T^T$ with $\lambda_i(X_\infty Y_\infty)$ remaining invariant. We can conceive of choosing a basis so that

$$X_\infty = Y_\infty = \mathrm{diag}(\nu_1, \nu_2, \ldots, \nu_n), \qquad \nu_i \geq \nu_{i+1}. \tag{4.4.8}$$

One can now conceive of truncating the plant realization $\{A, B, C\}$ of P and the controller realization $\{A_c, B_c, C_c\}$ of K (by discarding blocks associated with entries $\nu_{k+1}, \ldots, \nu_n$ of X_∞ and Y_∞). It is not hard to verify that the reduced order controller $\hat{K}$ stabilizes and achieves the closed-loop gain constraint for the reduced order plant $\hat{P}$ of the same dimension. (The key is to observe that the original Riccati equations immediately yield reduced order Riccati equations via the truncation process).

The key question is whether a reduced order controller $\hat{K}$ of some dimension will function satisfactorily with the original and (unreduced) plant P.

It turns out that a convenient way to analyze this problem is to work with *normalized coprime factors of a scaled plant.* More precisely, assume that

$$\gamma > \max(1, \gamma_0) \tag{4.4.9}$$

and set

$$\beta = \sqrt{1 - \gamma^{-2}}. \tag{4.4.10}$$

As explained in Mustafa and Glover (1991), it is only possible to have $\gamma_0 < 1$ if P is asymptotically stable with Hankel norm less than one. Under this circumstance, the theory we are presenting only has value when we choose $\gamma > 1$.

Let $\beta P = \bar{N}^{-1}(\beta\bar{M})$ represent βP using normalized left coprime factors. Using Y_∞, we can write

$$\begin{bmatrix} \beta\bar{M} & \bar{N} \end{bmatrix} = \left[\begin{array}{c|cc} A - \beta^2 Y_\infty C^T C & \beta B & -\beta^2 Y_\infty C^T \\ \hline C & 0 & I \end{array}\right]. \tag{4.4.11}$$

The controllability and observability grammians are $\bar{P} = \beta^2 Y_\infty$ and $\bar{Q} = (X_\infty^{-1} + \beta^2 Y_\infty)^{-1}$, which are both diagonal when $X_\infty = Y_\infty$ as in (4.4.8), and the Hankel singular values σ_l obey $\sigma_l^2 = \beta^2\nu_l^2(1 + \beta^2\nu_l^2)^{-1}$. Truncation of the original plant realization obtained by balancing the two Riccati equations (4.5.4) and (4.5.5) yields the same result as truncating a balanced realization of $[\beta\bar{M}\ \bar{N}]$. Suppose now that the truncation process results in

$$\beta\hat{\epsilon} := \left\| \left[\beta\left(\bar{M} - \hat{\bar{M}}\right) \quad \left(\bar{N} - \hat{\bar{N}}\right)\right] \right\|_\infty. \tag{4.4.12}$$

Suppose also that

$$\hat{\gamma} := \left\| \bar{T}\left(\hat{P}, \hat{K}\right) \right\|_\infty. \tag{4.4.13}$$

The key conclusions are:

$$\hat{\epsilon}(\beta + \hat{\gamma}) < 1 \tag{4.4.14}$$

ensures that $\hat{K}$ stabilizes P, and

$$\left\|\bar{T}(P, \hat{K})\right\|_\infty \leq \hat{\gamma} + \frac{\hat{\epsilon}(1+\hat{\gamma})(1+\beta+\hat{\gamma})}{1-\hat{\epsilon}(\beta+\hat{\gamma})}. \tag{4.4.15}$$

Further, there exists an overbound for $\hat{\epsilon}$ in terms of the Hankel singular values of $[\beta\bar{M}\ \bar{N}]$:

$$\hat{\epsilon} \leq 2\sum_{i=k+1}^{n} \frac{\nu_i}{\sqrt{1+\beta^2\nu_i^2}}, \tag{4.4.16}$$

when the reduction of K is to a $\hat{K}$ of order k. Also, $\hat{\gamma} < \gamma$. Consequently, a more conservative condition for stability is

$$\sum_{i=k+1}^{n} \frac{2\nu_i}{\sqrt{1+\beta^2\nu_i^2}}(\beta+\gamma) < 1 \tag{4.4.17}$$

and a more conservative bound on the closed-loop gain is

$$\left\|T(P, \hat{K})\right\|_\infty \leq \gamma + \frac{\sum_{i=k+1}^{n} \frac{2\nu_i}{\sqrt{1+\beta^2\nu_i^2}}(1+\gamma)(1+\beta+\gamma)}{1-\sum_{i=k+1}^{n} \frac{2\nu_i}{\sqrt{1+\beta^2\nu_i^2}}(\beta+\gamma)}. \tag{4.4.18}$$

It is well known that many LQG results can be obtained from H_∞ results by letting γ tend to ∞. This works as far as the Riccati equations are concerned. However, the stability condition (which is only a sufficient condition) can never be satisfied.

We now indicate briefly how the approach of Nagado *et al.* (1991) and Nagado *et al.* (1994) differ. The key is to work directly with the controller. A normalized right coprime realization $(\beta X)Y^{-1}$ is found for βK, and it is reduced. Define $\hat{\epsilon}_c$ by

$$\beta\hat{\epsilon}_c = \left\|\begin{matrix} Y-\hat{Y} \\ \beta(X-\hat{X})\end{matrix}\right\|_\infty \quad \text{and define} \quad \gamma_c = \left\|\bar{T}(P, K)\right\|_\infty.$$

A sufficient condition for closed-loop stability of the $(P, \hat{K})$ loop is

$$\hat{\epsilon}_c(\beta+\gamma_c) < 1$$

and

$$\left\|\bar{T}(P, \hat{K})\right\|_\infty \leq \gamma_c + \frac{\hat{\epsilon}_c(1+\gamma_c)(1+\beta+\gamma_c)}{1-\hat{\epsilon}_c(\beta+\gamma_c)}.$$

An example illustrates that for a range of gains, the Mustafa-Glover method provides higher bounds on the closed-loop gain and higher actual gain than the bounds and actual gain from the Nagado *et al.* method.

Main points of the section

1. Procedures can be given for reducing a controller for a restricted class of H_∞ problem. The key is to secure balanced solutions of the two H_∞ Riccati equations through coordinate basis change.

2. One method relies on obtaining a reduced order plant-controller pair which gives acceptable performance, and then using the controller on the original plant. *a priori* tests are available for the stability and performance of the closed-loop system containing the reduced order controller.

4.5 Controller Reduction via Coprime Fractions and Frequency Weighting

In this section, we shall explain another approach to reducing controllers, still based on coprime fractions, but now using frequency weighting. The ideas can be found in Liu, Anderson and Ly (1990). This paper includes a striking example on the reduction of a 55th order LQG designed controller for a 55th order unstable, nonminimum phase, two input, two output plant. The controller is to eliminate flutter on a B-767 aeroplane.

Basis of the algorithm

Suppose that $K(s) = \bar{Y}^{-1}(s)\bar{X}(s) = X(s)Y^{-1}(s)$ and $P(s) = \bar{N}^{-1}(s)\bar{M}(s) = M(s)N^{-1}(s)$ and suppose the double Bezout identity holds:

$$\begin{bmatrix} \bar{N} & -\bar{M} \\ \bar{X} & \bar{Y} \end{bmatrix} \begin{bmatrix} Y & M \\ -X & N \end{bmatrix} = \begin{bmatrix} I & 0 \\ 0 & I \end{bmatrix}. \tag{4.5.1}$$

Suppose further, with no loss of generality, that $\bar{Y}(\infty) = I$. As we know, the loop shown in Figure 4.5.1 is stable.

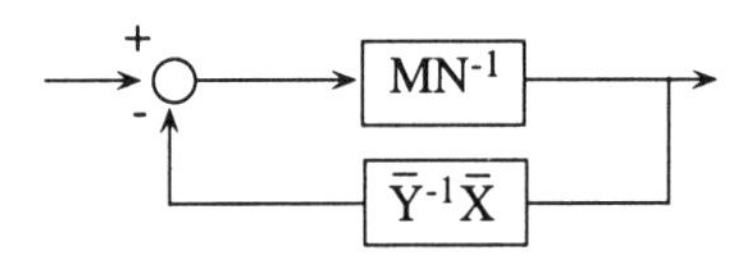

Figure 4.5.1. Closed loop in conventional form

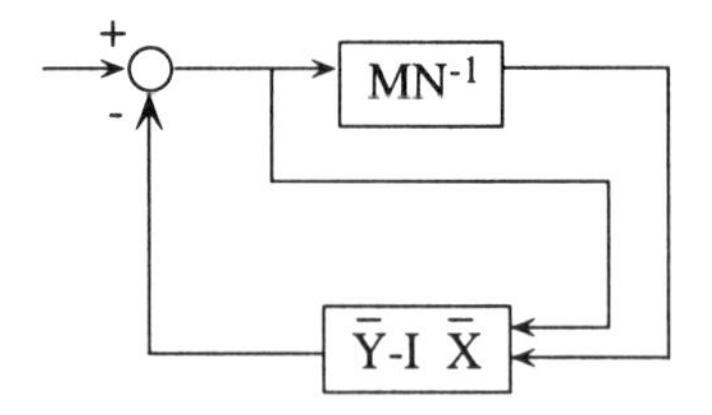

Figure 4.5.2. Redrawing of closed loop

It is trivial to see that, as a result, the loop shown in Figure 4.5.2 is also stable. The controller now has two (possibly vector) inputs, and strictly proper transfer function matrix $[\bar{Y} - I \;\; \bar{X}]$. Figure 4.5.3 shows a rewriting of that loop. Figure 4.5.3 depicts

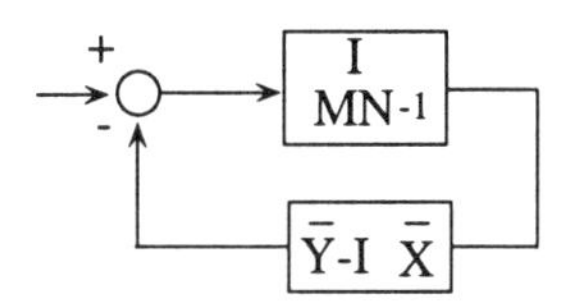

Figure 4.5.3. Further redrawing of closed loop

the same two (vector) input, single (vector) output controller, with a stable transfer function matrix. The plant is single (vector) input, two (vector) output. Denote the nonstandard plant and controller transfer function matrices appearing in Figure 4.5.3 by $P_a(s)$ and $K_a(s)$ (a = augmented). Now (following earlier ideas) seek a stable reduced order $\hat{K}_a(s)$ such that the measure

$$J = \left\| \left[K_a(s) - \hat{K}_a(s)\right] P_a(s) \left[I + K_a(s) P_a(s)\right]^{-1} \right\|_\infty \tag{4.5.2}$$

is as small as possible, and certainly less than 1 (which will guarantee stability when $\hat{K}_a(s)$ is introduced in place of $K_a(s)$). Let

$$\hat{K}_a(s) = \left[\hat{\bar{Y}} - I \quad \hat{\bar{X}}\right], \tag{4.5.3}$$

where it is reasonable to expect that $\hat{\bar{Y}} - I$ is strictly proper.

Then we shall take as the controller for the original plant $\hat{K}(s) = \hat{\bar{Y}}^{-1} \hat{\bar{X}}$. Of course, we want the degree of $[\bar{Y} \;\; \bar{X}]$ to be the same as that of $K(s)$, in order that the algorithm produces a $\hat{K}(s)$ of lesser degree than $K(s)$.

Now observe that

$$\begin{aligned} P_a[I + K_a P_a]^{-1} &= \begin{bmatrix} I \\ MN^{-1} \end{bmatrix} [I + \bar{Y} - I + \bar{X} M N^{-1}]^{-1} \\ &= \begin{bmatrix} N \\ M \end{bmatrix} [\bar{Y} N + \bar{X} M]^{-1} = \begin{bmatrix} N \\ M \end{bmatrix}. \end{aligned} \tag{4.5.4}$$

Hence the measure J of (4.5.2) is simply

$$J = \left\| \begin{bmatrix} \bar{Y} - \hat{\bar{Y}} & \bar{X} - \hat{\bar{X}} \end{bmatrix} \begin{bmatrix} N \\ M \end{bmatrix} \right\|_\infty . \tag{4.5.5}$$

By considering the 2-2 entry of (4.5.1), we can see that the approximation algorithm is trying to preserve satisfaction of part of the Bezout identity.

Of course, the optimization problem of minimizing J over $\begin{bmatrix} \hat{\bar{Y}} & \hat{\bar{X}} \end{bmatrix}$ of constrained degree is too hard to solve exactly. Instead, we can simply use frequency weighted balanced truncation, as described in the previous chapter.

To this end, it is helpful to observe how straightforward some of the calculations are in case $\bar{X}$, $\bar{Y}$, M, N are given by the earlier formulas of Section 4.2.

$$\begin{aligned} \bar{X} &= F(sI - A - HC)^{-1}H, \\ \bar{Y} &= I - F(sI - A - HC)^{-1}B, \\ M &= C(sI - A - BF)^{-1}B, \\ N &= I + F(sI - A - BF)^{-1}B. \end{aligned} \tag{4.5.6}$$

Then it is straightforward to verify that

$$\begin{bmatrix} \bar{Y} - I & \bar{X} \end{bmatrix} \begin{bmatrix} N \\ M \end{bmatrix} = \left[\begin{array}{cc|c} A + HC & -BF + HC & -B \\ 0 & A + BF & B \\ \hline F & 0 & 0 \end{array} \right]. \tag{4.5.7}$$

The controllability grammian can be verified to be

$$\begin{bmatrix} P & -P \\ -P & P \end{bmatrix},$$

where P satisfies the Lyapunov equation

$$(A + BF)P + P(A + BF)^T + BB^T = 0. \tag{4.5.8}$$

The observability grammian can be verified to be

$$\begin{bmatrix} Q & \star \\ \star & \star \end{bmatrix},$$

where Q satisfies the Lyapunov equation

$$Q(A + HC) + (A + HC)^T Q + F^T F = 0. \tag{4.5.9}$$

and the $\star$ entries are immaterial. The advantage comes from being able to compute the 1-1 block entry of the controllability grammian by solving a Lyapunov equation of size dictated by the plant alone. Normally, a larger size Lyapunov equation has to be solved, with dimension the sum of the plant order and the weight order.

Loop transfer recovery and connection with Enns' method

In this subsection, we shall show that when a controller is obtained by the so-called loop transfer recovery method (Anderson and Moore, 1989), the use of the scheme just presented and the use of the Enns (1984) scheme yield effectively the same result, *i.e.*, the two methods effectively coincide.

Loop transfer recovery it will be recalled has the aim of retaining for an output feedback LQG design the helpful robustness properties of an LQ design implemented with state feedback. It is achievable when the plant is square with all zeros in the open left half plane, and relies on a special choice for the process noise covariance matrix in the Kalman filter Riccati equation. With plant $C(sI - A)^{-1}B$, measurement noise covariance matrix V and unamended process noise covariance matrix W, the latter is adjusted to $W + q^2BB^T$ where q is a very large scalar. Thus the (filter) Riccati equation is

$$AS + SA^T - SC^TV^{-1}CS + W + q^2BB^T = 0 \tag{4.5.10}$$

and of course a stabilizing solution is sought. Given the assumptions on the plant it turns out that the Kalman filter gain, $-SC^TV^{-1}$, approaches a limit as q tends to ∞:

$$H \to -qBUV^{-1/2}, \tag{4.5.11}$$

where U is an orthogonal matrix depending on the parameters in the problem.

It is then possible to study the behaviour of $\bar{X}$ and $\bar{Y}$ in (4.5.6). The calculations are not very difficult and are set out in Liu and Anderson (1990). The conclusion is:

$$\begin{aligned} \bar{X}(s) &\to F(sI - A)^{-1}BP^{-1}(s), \\ \bar{Y}(s) &\to I. \end{aligned} \tag{4.5.12}$$

If we then study the reduction problem with weighting described in the previous subsection, see (4.5.5), it is evident that we are trying to minimize an index

$$J = \left\| \left(F(sI - A)^{-1}BNM^{-1} - \hat{\bar{X}} \right) M \right\|_\infty . \tag{4.5.13}$$

For the reduced order controller, $\hat{\bar{Y}} = I$ is the obvious approximation and accordingly we will set $\hat{K}(s) = \hat{\bar{X}}$. Now suppose we were instead to adopt the Enns approach in seeking to reduce the controller. Our task is to minimize an index

$$J = \left\| (K - \hat{K})P(I + KP)^{-1} \right\|_\infty . \tag{4.5.14}$$

By (4.5.12), we see that $K(s)$ approaches $F(sI - A)^{-1}BNM^{-1}$. Also, an easy calculation using the Bezout identity and the facts that $P = MN^{-1}$ and $K = \bar{Y}^{-1}\bar{X}$ yields that $P(I + KP)^{-1} = M\bar{Y} \to M$. Accordingly the Enns' index of (4.5.14) is actually

$$J = \left\| \left(F(sI - A)^{-1}BNM^{-1} - \hat{K}\right)M \right\|_{\infty}. \tag{4.5.15}$$

Since, as already mentioned, the quantity $\hat{\bar{X}}$ appearing in (4.5.13) is taken to be the same as $\hat{K}(s)$, *i.e.*, $\hat{\bar{Y}}$ is taken as I, we see that both approaches yield the same result.

Main points of the section

1. When the controller is described using a coprime fraction description, a frequency weighted reduction procedure can be found (based on a stability retention analysis). This procedure reduces the components of the coprime fraction.
2. When the controller has been obtained by loop transfer recovery, its reduction by this scheme is equivalent to using the basic Enns' scheme of frequency weighting balanced truncation.

4.6 Controller Reduction via Multiplicative Coprime Factor Approximation

In this section, we shall show how a relative error approximation approach can be used to find a reduced order controller, following an idea in Gu (1995). The starting point is a set of coprime descriptions for the plant $P(s) = \bar{N}^{-1}\bar{M} = MN^{-1}$ and the controller $K(s) = \bar{Y}^{-1}\bar{X} = XY^{-1}$ which also satisfy the double Bezout identity:

$$\begin{bmatrix} \bar{N} & -\bar{M} \\ \bar{X} & \bar{Y} \end{bmatrix} \begin{bmatrix} Y & M \\ -X & N \end{bmatrix} = I. \tag{4.6.1}$$

Suppose the plant $P(s)$ has a minimal state variable realization defined by the triple $\{A, B, C\}$. For an observer-based state feedback controller, there are F and H such that the eigenvalues of $A + BF$ and $A + HC$ lie in the open left half plane, and we can assume, see (4.2.18) and (4.2.21),

$$\begin{bmatrix} Y & M \\ -X & N \end{bmatrix} = \begin{bmatrix} I & 0 \\ 0 & I \end{bmatrix} + \begin{bmatrix} C \\ F \end{bmatrix} (sI - A - BF)^{-1} \begin{pmatrix} -H & B \end{pmatrix}, \tag{4.6.2}$$

so that the matrix on the left side of (4.6.2) has the same order as the plant itself. However, this is not necessary. For convenience, we shall rewrite the identity (4.6.1) in the following way.

$$\begin{bmatrix} \bar{N} & \bar{M} \\ -\bar{X} & \bar{Y} \end{bmatrix} \begin{bmatrix} Y & -M \\ X & N \end{bmatrix} = I. \tag{4.6.3}$$

The two square matrices appearing in the above product have entries with all poles in the open left half plane. Also, the fact that the identity holds implies that the inverse of each matrix has the same property. Thus each matrix is stable and minimum phase, and nonsingular on the extended imaginary axis. We can evidently attempt multiplicative approximation by the relative error method, *i.e.*, we can seek $\hat{Y}$, $\hat{X}$, $\hat{M}$ and $\hat{N}$ such that

$$\begin{bmatrix} \hat{Y} & -\hat{M} \\ \hat{X} & \hat{N} \end{bmatrix} = (I - \Delta) \begin{bmatrix} Y & -M \\ X & N \end{bmatrix}, \tag{4.6.4}$$

where $\Delta \in RH_\infty$ and we seek to minimize the H_∞ norm of Δ, subject to a degree constraint on the approximating matrix.

Notice that this equation implies

$$\begin{bmatrix} \hat{Y} \\ \hat{X} \end{bmatrix} = (I - \Delta) \begin{bmatrix} Y \\ X \end{bmatrix}. \tag{4.6.5}$$

We use the quantities $\hat{X}$ and $\hat{Y}$ to define the new controller, through $\hat{K} = \hat{X}\hat{Y}^{-1}$. The order of $\hat{K}$ will necessarily be at most the order of the transfer function matrix on the left side of (4.6.4). By contrast, the quantities $\hat{M}$ and $\hat{N}$ (which could be associated with a plant approximation) are not used; the intention is to use the reduced order controller in conjunction with the original plant.

The procedure just suggested is simple to understand. But is it worthwhile in terms of the quality of the approximation? We shall exhibit a nice approximation result, which serves to confirm the utility of the procedure. Recall the matrix $T(P, K)$, which captures a number of input-output properties of the closed loop and is given by

$$T(P, K) = \begin{bmatrix} I \\ K \end{bmatrix} [I + PK]^{-1} \begin{bmatrix} I & P \end{bmatrix}. \tag{4.6.6}$$

If we use the coprime fractional descriptions, this evaluates easily to

$$T(P, K) = \begin{bmatrix} Y \\ X \end{bmatrix} \begin{bmatrix} \bar{N} & \bar{M} \end{bmatrix}. \tag{4.6.7}$$

We also noted earlier (see Section 4.3) the following error formula, which can be derived using algebraic manipulations:

$$\begin{aligned} &T(P, \hat{K}) - T(P, K) \\ &= [I - T(P, K)] \begin{bmatrix} \hat{Y} - Y \\ \hat{X} - X \end{bmatrix} \begin{bmatrix} \bar{N} & \bar{M} \end{bmatrix} \left\{ I + \begin{bmatrix} \hat{Y} - Y \\ \hat{X} - X \end{bmatrix} \begin{bmatrix} \bar{N} & \bar{M} \end{bmatrix} \right\}^{-1}. \end{aligned} \tag{4.6.8}$$

In the light of (4.6.5) and (4.6.7), this can be rewritten as

$$T(P, \hat{K}) - T(P, K) = [I - T(P, K)](-\Delta T)[I - \Delta T]^{-1}. \tag{4.6.9}$$

Now we see that if $\|\Delta\|_\infty \|T\|_\infty < 1$, the reduced order controller is guaranteed to be stabilizing, and the error in closed-loop performance is bounded as follows:

$$\left\| T(P, \hat{K}) - T(P, K) \right\|_\infty \leq \frac{\|(P, K)\|_\infty^2 \|\Delta\|_\infty}{1 - \|\Delta\|_\infty \|(P, K)\|_\infty}. \tag{4.6.10}$$

(We are using the fact, noted earlier in Section 4.3, $\|T\|_\infty = \|I - T\|_\infty$.)

A point made clear by this analysis is that the bigger $\|T\|_\infty$ is, the less scope there is to approximate the controller. This fact is a general one: closed loop systems with large $\|T\|_\infty$ are known to be intolerant of changes of the controller (Vinnicombe, 1993).

Main points of the section

1. A relative error approximation can be used in a 2×2 block matrix containing coprime descriptions of the controller and the plant. A coprime description of a reduced order controller is obtained.

2. An error bound on closed-loop performance is available involving the original closed-loop transfer functions and the multiplicative error arising in the approximation step.

4.7 Reducing Controller Dimension to Preserve H_∞ Performance

In this section, we shall summarize a procedure for determining a low order controller from a full order controller, where the full order controller has been selected to fulfil an H_∞ design criterion. The method is quite different from the schemes based on Riccati equation solution balanced truncation, discussed earlier in this chapter. Here, there is no restrictions on the form of H_∞ problem, in contrast to the earlier situation.

We shall begin by highlighting certain results of the H_∞ theory; for a more complete discussion, see for example Zhou *et al.* (1996) and Green and Limebeer (1995).

The data is a plant $P(s)$, with two identified groups of inputs, disturbance inputs w and control inputs u, and two identified groups of outputs, to-be-controlled outputs z and measurable outputs for feedback, y:

$$\begin{bmatrix} z \\ y \end{bmatrix} = P(s) \begin{bmatrix} w \\ u \end{bmatrix}. \tag{4.7.1}$$

The aim is to find a controller

$$u = K(s)y, \tag{4.7.2}$$

such that the closed loop is stabilized and the resulting closed-loop transfer function matrix linking w to z has H_∞ gain less than some prescribed value γ.

It is usual to make a number of standing assumptions on $P(s)$ in order that the problem be solvable, see the references for details. The solution is then normally not unique; rather a set of acceptable controllers $K(s)$ is available, given by the following formula:

$$K(s) = \mathcal{F}_l(M, Q), \qquad Q \in RH_\infty, \qquad \|Q\|_\infty < \gamma. \tag{4.7.3}$$

Here, $M(s)$ is given as a result of calculations using $P(s)$ set out in the references by

$$M(s) = \begin{bmatrix} M_{11}(s) & M_{12}(s) \\ M_{21}(s) & M_{22}(s) \end{bmatrix} = \left[\begin{array}{c|cc} A & B_1 & B_2 \\ \hline C_1 & D_{11} & D_{12} \\ C_2 & D_{21} & D_{22} \end{array}\right], \tag{4.7.4}$$

with M_{11} having $\dim u$ rows and $\dim y$ columns, and M_{22} having $\dim y$ rows and $\dim u$ columns. The notation $\mathcal{F}_l(M, Q)$ means the transfer function matrix generated by connecting Q in feedback form from the second set of outputs of M to the second set of inputs, *i.e.*,

$$\mathcal{F}_l(M, Q) = M_{11} + M_{12}Q(I - M_{22}Q)^{-1}M_{21}. \tag{4.7.5}$$

Now define the matrix $\bar{\Theta}$ in terms of the state variable matrices making up M as follows:

$$\bar{\Theta} = \left[\begin{array}{c|cc} A - B_2D_{12}^{-1}C_1 & B_1 - B_2D_{12}^{-1}D_{11} & -B_2D_{12}^{-1} \\ \hline C_2 - D_{22}D_{21}^{-1}C_1 & D_{21} - D_{22}D_{12}^{-1}D_{11} & -D_{22}D_{12}^{-1} \\ D_{21}^{-1}C_1 & D_{22}D_{12}^{-1} & D_{21} - D_{22}D_{12}^{-1}D_{11} \end{array}\right]. \tag{4.7.6}$$

Some algebra will then show that the set of $K(s)$ solving the H_∞ problem can be described as

$$K(s) = \left(Q\bar{\Theta}_{12} + \bar{\Theta}_{22}\right)^{-1}\left(Q\bar{\Theta}_{11} + \bar{\Theta}_{21}\right) = \bar{Y}^{-1}(Q)\bar{X}(Q), \tag{4.7.7}$$

where $Q \in RH_\infty$ and $\|Q\|_\infty < \gamma$. The controller $K_0(s) = \bar{\Theta}_{22}^{-1}\bar{\Theta}_{21}$ obtained by taking $Q = 0$ is termed the central controller. Its order is normally that of the plant. *Our interest is in finding a Q obeying the restrictions and ensuring that the controller degree is reduced.* The way this can be achieved is set out in the following result of Zhou *et al.* (1996), which can be consulted for a proof. Suppose that

$$\left\| \begin{bmatrix} \bar{\Theta}_{21} - \hat{\bar{X}} & \bar{\Theta}_{22} - \hat{\bar{Y}} \end{bmatrix} \bar{\Theta}^{-1} \begin{bmatrix} \gamma^{-1}I & 0 \\ 0 & I \end{bmatrix} \right\|_\infty < \frac{1}{\sqrt{2}}. \tag{4.7.8}$$

Then $\hat{K}(s) = \hat{\bar{Y}}^{-1}(s)\hat{\bar{X}}(s)$ is also a stabilizing controller for $P(s)$ which achieves the H_∞ gain constraint.

What is then needed is simply a weighted reduction of the particular coprime factorization of the central controller, and of course the weighted error should respect the bound of (4.7.8). Note that the result being appealed to is a sufficiency result; there is no guarantee that if a reduced order controller exists which satisfies the H_∞ design constraint, such a controller can be found by the weighted approximation problems inherent in (4.7.8). For any approximants found using the index on the left of (4.7.8) may not give an error fulfilling the bound on the right side of (4.7.8).

Main points of the section

1. For controllers obtained as a result of a general H_∞ design, there exists a method for order reduction which seeks to preserve H_∞ performances. It depends on a frequency-weighted reduction of a particular coprime factorization of the central controller for the H_∞ problem.

2. The main result is a sufficiency result; it may be that satisfactory reduced order controllers exist, but the method will not find them.

4.8 Example

The controller reduction schemes presented in earlier sections are applied here to a numerical example. The plant is the same as that of Section 3.8, and the same high order controller designed by LQG techniques is used for the following several schemes.

First we consider the scheme of Section 4.3. We calculate a normalized coprime description $\bar{N}^{-1}(s)\bar{M}(s)$ of the plant, and then obtain a right coprime description $[X^T(s)\ Y^T(s)]^T$ of the controller such that $\bar{N}Y + \bar{M}X = I$; procedures are set out in Section 4.2. Next, $[X^T(s)\ Y^T(s)]^T$ is reduced to a fourth order $[\hat{X}^T(s)\ \hat{Y}^T(s)]^T$ by balanced truncation with no weighting. The reduced order controller is given by $\hat{K}(s) = \hat{X}(s)\hat{Y}^{-1}(s)$. Figure 4.8.1 shows the comparison of the closed loop system $P\hat{K}(I + P\hat{K})^{-1}$ with the original closed loop system $PK(I + PK)^{-1}$. The matching of gain characteristics from the low frequency region up to 0.6 rad/s is good. With the designed controller, the error norm $\|[\Delta_y^T\ \Delta_x^T]^T\|_\infty = 4.452\,419 \times 10^{-2}$. One can check that $\|T(P, K)\|_\infty = 117.2154$, and then the upper bound on $\|T(P, \hat{K}) - T(P, K)\|_\infty$ is 5.462 117, by (4.3.15). This is conservative: the actual value of $\|T(P, \hat{K}) - T(P, K)\|_\infty$ is $3.201\,850 \times 10^{-1}$.

Second, let us apply multiplicative approximation of coprime factors to the same plant and controller, as in Section 4.6. The coprime description $[X^T(s)\ Y^T(s)]^T$ of the controller is reduced to the fourth order one $[\hat{X}_m^T(s)\ \hat{Y}_m^T(s)]^T$ through the multiplicative error defined by (4.6.5). The reduced order controller $\hat{K}_m(s) = \hat{X}_m(s)\hat{Y}_m^{-1}(s)$ is applied to the original system. The closed loop gain characteristic $P\hat{K}_m(I + P\hat{K}_m)^{-1}$ is also shown in Figure 4.8.1. The result for $\hat{K}(s)$ is similar to the result for $\hat{K}_{wbt}(s)$ in Section 3.8, while $\hat{K}_m(s)$ gives superior results except for a narrow band around 0.75 rad/s.

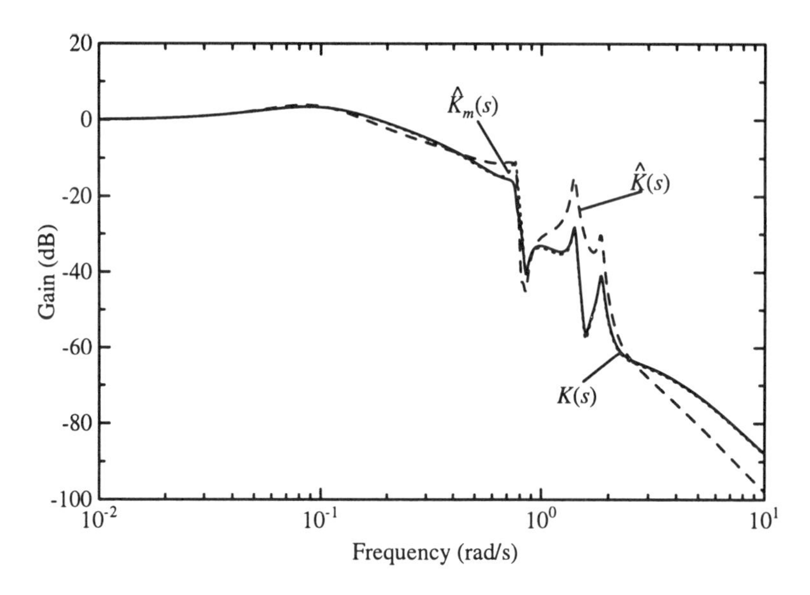

Figure 4.8.1. Comparison of closed loop transfer functions with original controller and two reduced order controllers

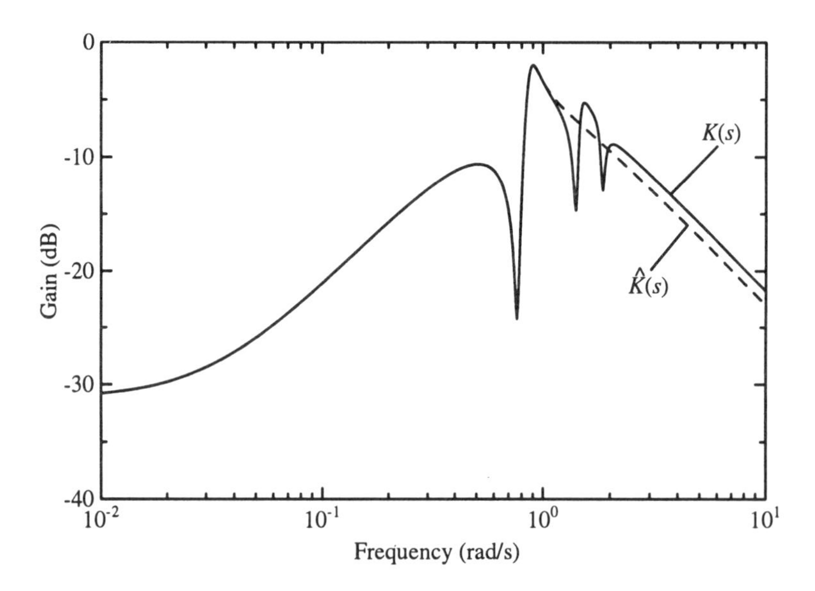

Figure 4.8.2. Comparison of controller gain characteristics

Let us examine how to retain H_∞ performance in controller reduction by the method of Section 4.7. The basic plant is still the four disk system, but this must be embedded in a generalized plant. The setup for the generalized plant is given as follows, see Zhou *et al.* (1996):

$$\dot{x} = \begin{bmatrix} A_1 \\ \begin{bmatrix} I_7 & 0_{7\times 1} \end{bmatrix} \end{bmatrix} x + B_1 w + B_2 u,$$
$$z = \begin{bmatrix} \sqrt{q_1} H \\ 0 \end{bmatrix} x + \begin{bmatrix} 0 & 1 \end{bmatrix}^T u, \qquad y = C_2 x + \begin{bmatrix} 0 & 1 \end{bmatrix} w, \tag{4.8.1}$$

$$\begin{aligned}
A_1 &= \begin{bmatrix} -0.161 & -6.004 & -0.58215 & -9.9835 & -0.40727 & -3.982 & 0 & 0 \end{bmatrix}, \\
B_2^T &= \begin{bmatrix} 1 & 0 & 0 & 0 & 0 & 0 & 0 & 0 \end{bmatrix}, \\
B_1 &= \begin{bmatrix} \sqrt{q_2} B_2 & 0 \end{bmatrix}, \\
H &= \begin{bmatrix} 0 & 0 & 0 & 0 & 0.55 & 11 & 1.32 & 18 \end{bmatrix}, \\
C_2 &= \Big[0 \quad 0 \quad 6.4432 \times 10^{-3} \quad 2.3196 \times 10^{-3} \\
&\qquad 7.1252 \times 10^{-2} \quad 1.0002 \quad 0.10455 \quad 0.99551 \Big],
\end{aligned}$$

where $q_1 = 1 \times 10^{-6}$, $q_2 = 1$, and w is the disturbance, z is the controlled variable. The controller $K(s)$ feeds back y to u. The transfer function from w to z is denoted by $T_{zw}(s)$. The infimum of $\|T_{zw}(s)\|_\infty$ over all stabilizing controller is 1.1272. Now, obtain a central controller $K(s)$ which gives the performance $\|T_{zw}(s)\|_\infty < 1.2$. The gain plot is shown in Figure 4.8.2. Calculate the coprime description $\bar{\Theta}_{22}^{-1}\bar{\Theta}_{21}$ of $K(s)$, and then set the index J to be minimized as follows:

$$\left\| \begin{bmatrix} \bar{\Theta}_{21} - \hat{\hat{X}} & \bar{\Theta}_{22} - \hat{\hat{Y}} \end{bmatrix} \bar{\Theta}^{-1} \begin{bmatrix} (1.2)^{-1} I & 0 \\ 0 & I \end{bmatrix} \right\|_\infty .$$

Here $\hat{\hat{X}}$ and $\hat{\hat{Y}}$ are coprime factors of a reduced order controller and $\bar{\Theta}$ is given by (4.7.6). We apply frequency weighted balanced truncation for approximately minimizing J. The fourth order pair of $(\hat{\hat{X}}, \hat{\hat{Y}})$ is obtained by use of frequency weighted balanced truncation, and then the fourth order controller is given as $\hat{K} = \hat{\hat{Y}}^{-1}\hat{\hat{X}}$. Figure 4.8.2 shows the comparison of $K(s)$ and $\hat{K}(s)$ in gain plots. The closed loop transfer function of the generalized plant (4.8.1) with the reduced order controller is denoted by $\hat{T}_{zw}(s)$. The four transfer function entries of $T_{zw}(s)$ and $\hat{T}_{zw}(s)$ are compared in Figures 4.8.3, 4.8.4, 4.8.5 and 4.8.6. The results of the controller reduction are evidently good. However, the value of J is 1.81439, so that the sufficiency condition (4.7.6) is not satisfied in this case. Nevertheless, the exact value of $\|\hat{T}_{zw}(s)\|_\infty$ is 1.1977; so, the H_∞ performance is good in comparison with the value of $\|T_{zw}(s)\|_\infty$.

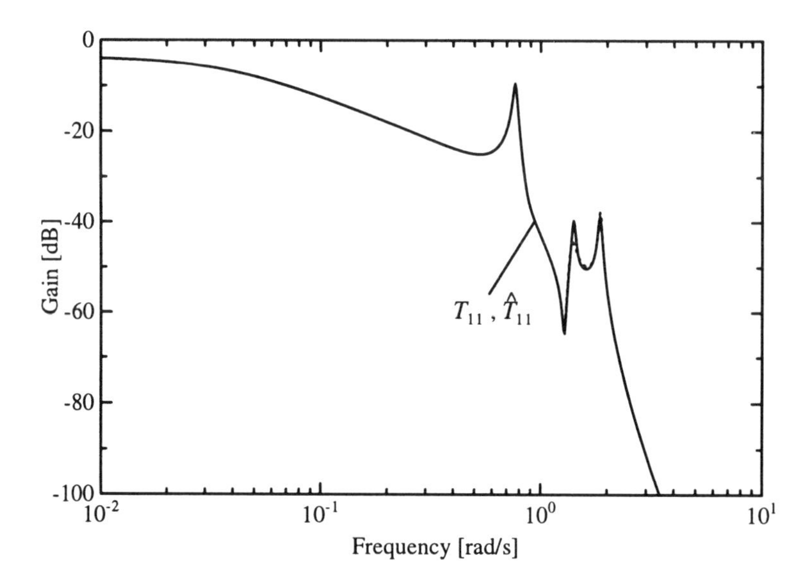

Figure 4.8.3. Gain plots of transfer function from w_1 to z_1

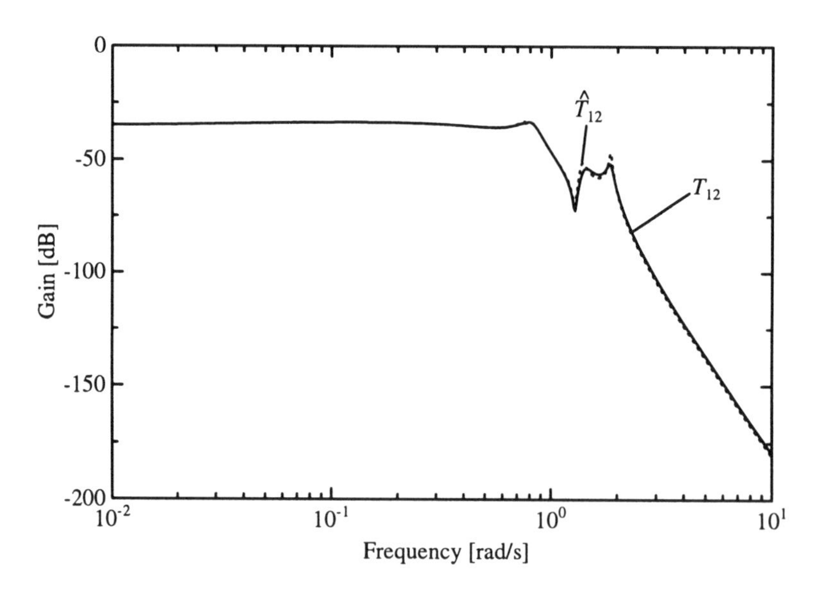

Figure 4.8.4. Gain plots of transfer function from w_2 to z_1

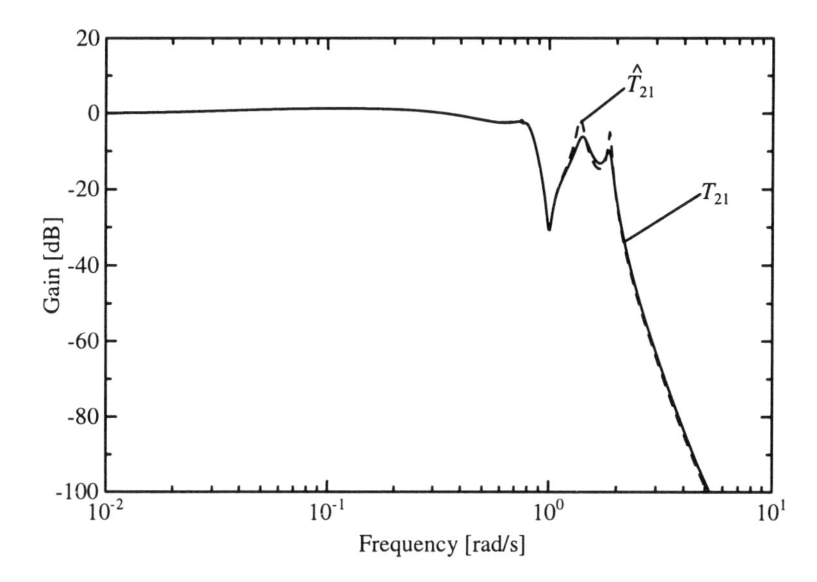

Figure 4.8.5. Gain plots of transfer function from w_1 to z_2

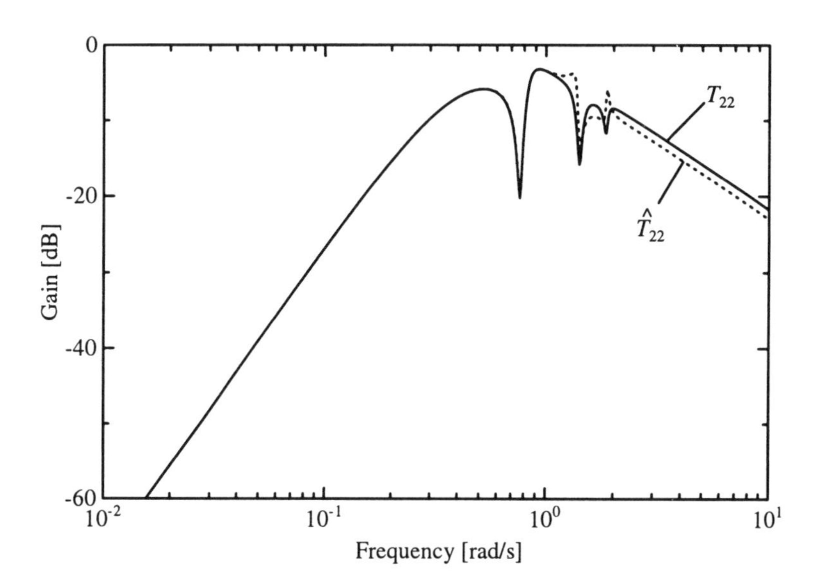

Figure 4.8.6. Gain plots of transfer function from w_2 to z_2

Bibliography

Al-Saggaf, U. M. and Franklin, G. F. (1986). On model reduction, *Proc. 25th IEEE Conf. on Decision and Control*, pp. 1064–1069.

Al-Saggaf, U. M. and Franklin, G. F. (1988). Model reduction via balanced realization: an extension and frequency weighted techniques, *IEEE Trans. on Automatic Control* **AC-33**: 687–692.

Anderson, B. D. O. (1986). Weighted Hankel norm approximation: calculation of bounds, *Systems & Control Letters* **7**: 247–255.

Anderson, B. D. O. and Liu, Y. (1989). Controller reduction: concepts and approaches, *IEEE Trans. on Automatic Control* **AC-34**: 802–812.

Anderson, B. D. O. and Moore, J. B. (1989). *Optimal Control: Linear Quadratic Methods*, Prentice-Hall, Englewood Cliffs, N.J.

Antoulas, A. C. and Bishop, R. H. (1987). Continued-fraction decomposition of linear systems in the state space, *Systems & Control Letters* **9**: 43–53.

Aoki, M. (1968). Control of large-scale dynamic systems by aggregation, *IEEE Trans. on Automatic Control* **AC-13**: 1–2.

Bernstein, D. S. and Hyland, D. C. (1985). The optimal projection equations for fixed-order dynamic compensation, *IEEE Trans. on Automatic Control* **AC-29**: 1034–1037.

Bonvin, D. and Mellichamp, D. A. (1982). A unified derivation and critical review of modal approaches to model reduction, *International J. of Control* **35**: 829–825.

Calfe, M. R. and Healey, M. (1974). Continued-fraction model-reduction technique for multivariable systems, *Proc. IEE* **121**: 393–395.

Chidambara, M. R. (1971). A new canonical form of state-variable equations and its application in the determination of a mathematical model of an unknown system, *International J. of Control* **14**: 897–909.

de Villemagne, C. and Skelton, R. E. (1987). Model reduction using a projection formulation, *International J. of Control* **46**: 2141–2169.

Desai, U. B. and Pal, D. (1984). A transformation approach to stochastic model reduction, *IEEE Trans. on Automatic Control* **AC-29**: 1097–1100.

Enns, D. (1984). Model reduction with balanced realizations: An error bound and a frequency weighted generalization, *Proc. 23rd IEEE Conf. on Decision and Control*, pp. 127–132.

Enright, W. and Kamel, M. (1980). On selecting a low–order model using the dominant mode concept, *IEEE Trans. on Automatic Control* **AC-25**: 976–978.

Gangsaas, D., Bruce, K. R., Blight, J. D. and Ly, U.-L. (1986). Application of modern synthesis to aircraft control: three case studies, *IEEE Trans. on Automatic Control* **AC-31**: 995–1104.

Georgiou, T. T. and Smith, M. C. (1990). Optimal robustness in the gap metric, *IEEE Trans. on Automatic Control* **AC-35**: 673–686.

Glover, K. (1984). All optimal Hankel-norm approximations of linear multivariable systems and their L^{∞} error bounds, *International Jounal of Control* **39**: 1115–1193.

Glover, K. (1986). Multiplicative approximation of linear multivariable systems with L^{∞} error bounds, *Proc. American Control Conference*, pp. 1705–1709.

Glover, K., Curtain, R. F. and Partington, J. R. (1988). Realization and approximation of linear infinite-dimensional systems and their L^{∞} - error bounds, *IEEE Trans. on Automatic Control* **AC-26**: 863–898.

Glover, K. and Limebeer, D. J. N. (1983). Robust multivariable control system design using optimal reduced order plant methods, *Proc. American Control Conf.* pp. 644–649.

Goddard, P. J. and Glover, K. (1993). Controller reduction: weights for stability and performance preservation, *Proc. 32nd IEEE Conf. on Decision and Control*, pp. 2903–2908.

Greeley, S. W. and Hyland, D. C. (1987). Reduced order compensation: LQG reduction versus optimal projection using a homotopic continuation method, *Proc. 26th IEEE Conf. on Decision and Control*, pp. 742–747.

Green, M. (1988a). Balanced stochastic realizations, *Linear Algebra and Applications* **98**: 211–247.

Green, M. (1988b). A relative-error bound for balanced stochastic truncation, *IEEE Trans. on Automatic Control* **AC-33**: 961–965.

Green, M. and Anderson, B. D. O. (1990). Generalized balanced stochastic truncation, *Proc. 29th IEEE Conf. on Decision and Control*, pp. 476–481.

Green, M. and Limebeer, D. J. N. (1995). *Linear Robust Control*, Prentice-Hall, Englewood Cliffs, N.J.

Gu, G. (1995). Model reduction with relative/multiplicative error bounds and relation to controller reduction, *IEEE Trans. on Automatic Control* **AC-40**: 1478–1485.

Helmersson, A. (1994). Model reduction using LMIs, *Proc. 33rd IEEE Conf. on Decision and Control*, pp. 3217–3222.

Horiguchi, K., Nishimura, T. and Nagata, A. (1990). Minimal realizatins interpolating first-and second-order information, *International J. of Control* **52**: 389–704.

Horn, R. A. and Johnson, C. R. (1991). *Topics in Matrix Analysis*, Cambridge University Press, Cambridge, U.K.

Hung, S. and Glover, K. (1985). Optimal Hankel-norm approximation of stable systems with first-order stable weighting functions, *Systems & Control Letters* **7**: 165–172.

Inooka, H. and Obinata, G. (1977). Mixed method of aggregation and I.S.E. criterion approaches for system reduction, *Electronics Letters* **13-3**: 88–90.

Inouye, Y. (1983). Approximation of multivariable linear systems with impulse response and autocorrelation sequences, *Automatica* **19**: 265–277.

Jonckheere, E. A. and Silverman, L. M. (1983). A new set of invariants for linear-systems-application to reduced order compensation design, *Systems & Control Letters* **AC-28**: 953–964.

Kabamba, P. T. (1985). Balanced gains and their significance for L_2 model reduction, *IEEE Trans. on Automatic Control* **AC-30**: 690–692.

Kim, S. W., Anderson, B. D. O. and Madievski, A. G. (1995a). Error bound for transfer function order reduction using frequency weighted balanced truncation, *Systems & Control Letters* **24**: 183–192.

Kim, S. W., Anderson, B. D. O. and Madievski, A. G. (1995b). Multiplicative approximation of transfer functions with frequency weighting, *Systems & Control Letters* **25**: 199–204.

Kimura, H. (1983). Optimal L_2-approximation with fixed poles, *Systems & Control Letters* **2**: 257–261.

Kokotovic, P., O'Malley, R. and Sannuti, P. (1976). Singular perturbations and order reduction in control theory – an overview, *Automatica* **12**: 123–132.

Kootsookos, P. J., Bitmead, R. R. and Green, M. (1992). The Nehari shuffle: FIR(q) filter design with guaranteed error bounds, *IEEE Trans. on Signal Processing* **40**: 1876–1883.

Lam, J. and Anderson, B. D. O. (1992). L_1 impulse response error bound for balanced truncation, *Systems & Control Letters* **18**: 129–137.

Latham, G. A. and Anderson, B. D. O. (1985). Frequency-weighted optimal Hankel norm approximation of stable transfer functions, *Systems & Control Letters* **5**: 229–236.

Lin, C.-A. and Chiu, T.-Y. (1992). Model reduction via frequency weighted balanced realization, *Control Theory and Advanced Technology* **8**: 341–351.

Liu, Y. and Anderson, B. D. O. (1989). Singular perturbation approximation of balanced systems, *International Journal of Control* **50**: 1379–1405.

Liu, Y. and Anderson, B. D. O. (1990). Frequency weighted controller reduction methods and loop transfer recovery, *Automatica* **23**: 487–497.

Liu, Y., Anderson, B. D. O. and Ly, U.-L. (1990). Coprime factorization controller reduction with Bezout identity induced frequency weighting, *Automatica* **26**: 233–249.

Madievski, A. G. and Anderson, B. D. O. (1995). Sampled-data controller reduction procedure, *IEEE Trans. on Automatic Control* **AC-40**: 1922–1926.

Matson, J. B., Lam, J., Anderson, B. D. O. and James, B. (1993). Multiplicative Hankel norm approximation of linear multivariable systems, *International J. of Control* **58**: 129–167.

Meier, L. and Luenberger, D. G. (1967). Approximation of linear constant systems, *IEEE Trans. on Automatic Control* **AC-12**: 585–588.

Meyer, D. G. (1990). Fractional balanced reduction: model reduction by fractional representation, *IEEE Trans. on Automatic Control* **AC-35**: 1341–1345.

Moore, B. C. (1981). Principal component analysis in linear systems: controllability, observability and model reduction, *IEEE Trans. on Automatic Control* **AC-26**: 17–31.

Mullis, C. T. and Roberts, R. A. (1976). The use of second-order information in the approximation of discrete-time systems, *IEEE Trans. on Acoustic Speech and Signal Processing* **ASSP-24**: 226–237.

Mustafa, D. and Glover, K. (1991). Controller reduction by H_∞ -balanced truncation, *IEEE Trans. on Automatic Control* **AC-36**: 668–682.

Nagado, T., Shimemura, E., Ishida, T. and Uchida, K. (1991). Controller reduction by an indirect method using Riccati equations, *Proc. 14th Dynamical Systems Theory Symposium*, pp. 207–210. in Japanese.

Nagado, T., Shimemura, E., Ishida, T. and Uchida, K. (1994). Controller reduction by indirect method using Riccati equations, *Trans of SICE*, Vol. 30, pp. 1266–1268. in Japanese.

Nett, C. N., Jacobson, C. A. and Balas, M. J. (1984). A connection between state-space and doubly coprime fractional representations, *IEEE Trans. on Automatic Control* **AC-29**: 831–832.

Obinata, G. (1978). Remarks on the time moments of composite systems, *Electronics Letters* **14–16**: 509–511.

Obinata, G. (1989). Low order models and robust control, *Measurement and Control* **28**: 988–995. in Japanese.

Obinata, G. and Inooka, H. (1976). A method for modeling linear time-invariant systems by linear systems of low order, *IEEE Trans. on Automatic Control* **AC-21**: 602–603.

Obinata, G. and Inooka, H. (1983). Author's reply to 'Comments on model reduction by minimizing the equation error', *IEEE Trans. on Automatic Control* **AC-28**: 124–125.

Obinata, G., Nakamura, T. and Inooka, H. (1988). An equation error method for the design of digital controllers, *IEEE Trans. on Automatic Control* **AC-33**: 384–386.

Pernebo, L. and Silverman, L. (1982). Model reduction via balanced state space representations, *IEEE Trans. on Automatic Control* **AC-27**: 382–387.

Rao, S. V. and Lamba, S. S. (1975). Eigenvalue assignment in linear optimal control systems via reduced-order models, *Proc. IEE* **122**: 197–201.

Safonov, M. and Chiang, R. (1989). A Schur method for balanced-truncation model reduction, *IEEE Trans. on Automatic Control* **AC-43**: 729–733.

Safonov, M. G. (1987). Imaginary-axis zeros in multivariable H_∞ optimal control, *in* R. F. Curtain (ed.), *Modeling, Robustness and Sensitivity Reduction in Control*, Springer-Verlag, Berlin.

Safonov, M. G. and Chiang, R. Y. (1988). Model reduction for robust control: a Schur relative error method, *International J. Adaptive Control and Signal Proc.* **2**: 259–272.

Seto, K. and Mitsuta, S. (1991). A new method for making a reduced-order model of flexible structures using unobservability and uncontrollability with application to vibration control, *Transactions Japan Society Mechanical Engineering* **57**: 3393–3399. in Japanese.

Shamash, Y. (1975a). Model reduction using minimal realisation algorithms, *Electronics Letters* **11-7**: 385–387.

Shamash, Y. (1975b). Model reduction using the Routh stability criterion and the *Padé* approximation technique, *International J. of Control* **21**: 475–484.

Siret, J., Michailesco, G. and Bertrand, P. (1977). Representation of linear dynamical systems by aggregated models, *International Journal of Control* **26**: 976–978.

Sreeram, V. and Anderson, B. D. O. (1995). Frequency weighted balanced reduction technique: a generalization and an error bound, *Proc. 34th IEEE Conf. on Decision and Control*, pp. 3576–3581.

Sreeram, V., Anderson, B. D. O. and Madievski, A. G. (1995). New results on frequency weighted balanced reduction technique, *Proc. American Control Conf.* pp. 4004–4009.

Sufleta, Z. (1984). A method for linear model reduction using the input-independent equation error, *Systems & Control Letters* **4**: 287–291.

Vidyasagar, M. (1985). *Control Systems Synthesis: A Factorization Approach*, MIT Press, Cambridge, MA.

Villemagne, de, C. (1987). *see* de Villemagne.

Vinnicombe, G. (1993). Frequency domain uncertainty and the graph topology, *IEEE Trans. on Automatic Control* **AC-38**: 1371–1383.

Wang, W. and Safonov, M. G. (1990). A tighter relative-error bound for balanced stochastic truncation, *Systems & Control Letters* **14**: 307–314.

Wang, W. and Safonov, M. G. (1991). Relative error bound for discrete balanced truncation, *International J. of Control* **54**: 593–612.

Wang, W. and Safonov, M. G. (1992). Multiplicative-error bound for balanced stochastic truncation model reduction, *IEEE Trans. on Automatic Control* **AC-37**: 1265–1267.

Wilson, D. A. (1970). Optimum solution of model-reduction problem, *Proc. IEE* **117**: 1161–1165.

Wright, D. J. (1973). The continued fraction representation of transfer functions and model simplification, *International J. of Control* **18**: 449–454.

Yousuff, A., Wagie, D. A. and Skelton, R. E. (1985). Linear system approximation via covariance equivalent realizations, *Journal Mathematical Analysis and Applications* **106**: 91–115.

Zakian, V. (1973). Simplification of linear-time invariant systems by moment approximants, *International J. of Control* **18**: 455–460.

Zhou, K. (1993a). Frequency weighted model reduction with L_∞ error bounds, *Systems & Control Letters* **21**: 115–125.

Zhou, K. (1993b). Weighted optimal Hankel norm model reduction, *Proc. 32nd IEEE Conf. on Decision and Control*, pp. 3353–3354.

Zhou, K. (1995). Frequency-weighted L_∞ norm and optimal Hankel norm model reduction, *IEEE Trans. on Automatic Control* **AC-40**: 1687–1699.

Zhou, K. and Chen, J. (1995). Performance bounds for coprime factor controller reductions, *Systems & Control Letters* **26**: 119–127.

Zhou, K., Doyle, J. C. and Glover, K. (1996). *Robust and Optimal Control*, Prentice-Hall, Englewood Cliffs, N.J.

Index